Stem Cell and Biologic Scaffold Engineering

Stem Cell and Biologic Scaffold Engineering

Special Issue Editor
Panagiotis Mallis

MDPI • Basel • Beijing • Wuhan • Barcelona • Belgrade

Special Issue Editor
Panagiotis Mallis
Biomedical Research Foundation Academy
of Athens (BRFAA),
Greece

Editorial Office
MDPI
St. Alban-Anlage 66
4052 Basel, Switzerland

This is a reprint of articles from the Special Issue published online in the open access journal *Bioengineering* (ISSN 2306-5354) from 2018 to 2019 (available at: https://www.mdpi.com/journal/bioengineering/special_issues/stem_cell_scaffold).

For citation purposes, cite each article independently as indicated on the article page online and as indicated below:

LastName, A.A.; LastName, B.B.; LastName, C.C. Article Title. *Journal Name* **Year**, *Article Number*, Page Range.

ISBN 978-3-03921-497-6 (Pbk)
ISBN 978-3-03921-498-3 (PDF)

Cover image courtesy of Panagiotis Mallis

© 2019 by the authors. Articles in this book are Open Access and distributed under the Creative Commons Attribution (CC BY) license, which allows users to download, copy and build upon published articles, as long as the author and publisher are properly credited, which ensures maximum dissemination and a wider impact of our publications.

The book as a whole is distributed by MDPI under the terms and conditions of the Creative Commons license CC BY-NC-ND.

Contents

About the Special Issue Editor vii

Preface to "Stem Cell and Biologic Scaffold Engineering" ix

Panagiotis Mallis, Catherine Stavropoulos-Giokas and Efstathios Michalopoulos
Introduction to the Special Issue on Stem Cell and Biologic Scaffold Engineering
Reprinted from: *Bioengineering* **2019**, *6*, 72, doi:10.3390/bioengineering6030072 1

Tiago P. Dias, Tiago G. Fernandes, Maria Margarida Diogo and Joaquim M. S. Cabral
Multifactorial Modeling Reveals a Dominant Role of Wnt Signaling in Lineage Commitment of Human Pluripotent Stem Cells
Reprinted from: *Bioengineering* **2019**, *6*, 71, doi:10.3390/bioengineering6030071 4

Panagiotis Mallis, Ioanna Gontika, Zetta Dimou, Effrosyni Panagouli, Jerome Zoidakis, Manousos Makridakis, Antonia Vlahou, Eleni Georgiou, Vasiliki Gkioka, Catherine Stavropoulos-Giokas and Efstathios Michalopoulos
Short Term Results of Fibrin Gel Obtained from Cord Blood Units: A Preliminary in Vitro Study
Reprinted from: *Bioengineering* **2019**, *6*, 66, doi:10.3390/bioengineering6030066 23

Vassilis Protogerou, Efstathios Michalopoulos, Panagiotis Mallis, Ioanna Gontika, Zetta Dimou, Christos Liakouras, Catherine Stavropoulos-Giokas, Nikolaos Kostakopoulos, Michael Chrisofos and Charalampos Deliveliotis
Administration of Adipose Derived Mesenchymal Stem Cells and Platelet Lysate in Erectile Dysfunction: A Single Center Pilot Study
Reprinted from: *Bioengineering* **2019**, *6*, 21, doi:10.3390/bioengineering6010021 37

Haobo Yuan
Introducing the Language of "Relativity" for New Scaffold Categorization
Reprinted from: *Bioengineering* **2019**, *6*, 20, doi:10.3390/bioengineering6010020 50

Panagiotis Mallis, Panagiota Chachlaki, Michalis Katsimpoulas, Catherine Stavropoulos-Giokas and Efstathios Michalopoulos
Optimization of Decellularization Procedure in Rat Esophagus for Possible Development of a Tissue Engineered Construct
Reprinted from: *Bioengineering* **2019**, *6*, 3, doi:10.3390/bioengineering6010003 60

Ioanna Gontika, Michalis Katsimpoulas, Efstathios Antoniou, Alkiviadis Kostakis, Catherine Stavropoulos-Giokas and Efstathios Michalopoulos
Decellularized Human Umbilical Artery Used as Nerve Conduit
Reprinted from: *Bioengineering* **2018**, *5*, 100, doi:10.3390/bioengineering5040100 70

Panagiotis Mallis, Dimitra Boulari, Efstathios Michalopoulos, Amalia Dinou, Maria Spyropoulou-Vlachou and Catherine Stavropoulos-Giokas
Evaluation of HLA-G Expression in Multipotent Mesenchymal Stromal Cells Derived from Vitrified Wharton's Jelly Tissue
Reprinted from: *Bioengineering* **2018**, *5*, 95, doi:10.3390/bioengineering5040095 82

About the Special Issue Editor

Panagiotis Mallis, M.Sc., Ph.D., is Research Associate at the Hellenic Cord Blood Bank of Biomedical Research Foundation Academy of Athens, Greece. His background is in Biomedical Sciences, and his Ph.D. focused on the development of small diameter vascular grafts utilizing the decellularized human umbilical artery. In recent years, he has focused on the publication of scientific articles and participated in conferences and presented invited talks. Panagiotis Mallis has extensive experience in mesenchymal stromal cell isolation and in vitro manipulation, where he studies their immunoregulatory/immunosuppressive properties and their applicability in tissue engineering and regenerative medicine approaches. His current research is mainly focused on the proper development of biological scaffolds—utilizing mostly decellularization methods—and their efficient combination with stem cells.

Preface to "Stem Cell and Biologic Scaffold Engineering"

Tissue engineering aims to achieve the restoration or substitution of damaged tissues and organs. The utilization of tissue engineering strategies has attracted the attention of the scientific community, and represents one of the highly emerging research fields of the 21st century.

Tissue engineering approaches involve the use of appropriate cellular populations combined with specific scaffolds, thus enabling cell adhesion, proliferation, and differentiation. The repopulated scaffolds can be implanted at sites of injury, thus expressing their regenerative properties. Nowadays, the design and in vitro development of whole organs is the primary focus. This new era of biotechnology will be of significant importance in the coming years for promoting personalized medicine. In this context, personalized medicine aims to improve the quality of life of patients. The ultimate goal of this effort is to make personalized regenerative medicine easily accessible to patients, regardless of their socioeconomic status.

This Special Issue emphasizes state-of-the-art tissue engineering and regenerative medicine approaches that consider the use of stem cells and biologically based scaffolds in order to achieve clinical utility. The current issue includes:

- Tissue engineering methods for the development of biologically based scaffolds;
- Biologically based scaffold implantation in animal models;
- Models establishing the regenerative properties of stem cells;
- Models for understanding the intracellular signaling pathways in induced pluripotent stem cells;
- Scientific opinions regarding scaffold categorization.

Panagiotis Mallis
Special Issue Editor

Editorial

Introduction to the Special Issue on Stem Cell and Biologic Scaffold Engineering

Panagiotis Mallis *, Catherine Stavropoulos-Giokas and Efstathios Michalopoulos

Hellenic Cord Blood Bank, Biomedical Research Foundation Academy of Athens, 4 Soranou Ephessiou Street, Athens 115 27, Greece
* Correspondence: pmallis@bioacademy.gr; Tel: +302106597340; fax: +30 210 6597345

Received: 17 August 2019; Accepted: 19 August 2019; Published: 21 August 2019

Abstract: Tissue engineering and regenerative medicine is a rapidly evolving research field that effectively combines stem cells and biologic scaffolds in order to replace damaged tissues. Biologic scaffolds can be produced through the removal of resident cellular populations using several tissue engineering approaches, such as the decellularization method. In addition, tissue engineering requires the interaction of biologic scaffolds with cellular populations. Stem cells are characterized by unlimited cell division, self-renewal, and differentiation potential, distinguishing themselves as a frontline source for the repopulation of decellularized matrices and scaffolds. However, parameters such as stem cell number, in vitro cultivation conditions, and specific growth media composition need further evaluation. The ultimate goal is the development of "artificial" tissues similar to native ones, which is achieved by properly combining stem cells and biologic scaffolds, thus bringing artificial tissues one step closer to personalized medicine. In this special issue of *Bioengineering*, we highlight the beneficial effects of stem cells and scaffolds in the emerging field of tissue engineering. The current issue includes articles regarding the use of stem cells in tissue engineering approaches and the proper production of biologically based scaffolds like nerve conduit, esophageal scaffold, and fibrin gel.

Keywords: tissue engineering; regenerative medicine; stem cells; scaffolds; MSCs; iPSCs; nerve conduit; fibrin gel; scaffold classification; Wnt signaling

---

Tissue engineering (TE) compromises an emerging field of the 21st century, where the repair or substitution of damaged tissues is still a clinical challenge. For the first time, in 1993, Langer and Vacanti proposed the definition of TE as "an interdisciplinary field that applies the principles of engineering and life sciences toward the development of biological substitutes that restore, maintain, or improve tissue function or a whole organ" [1]. For this purpose, TE could potentially be used in various regenerative medicine (RM) approaches, by efficiently combining stem cells and scaffolds.

In this context, stem cells can be classified into embryonic and adult stem cells [2]. Embryonic stem cells (ESCs), which are referred to as pluripotent stem cells, can give rise to any cell type or tissue, and can be derived from early embryo stages, like blastocyst and inner cell mass [2]. Due to ethical concerns, the use of ESCs is limited in TE and RM approaches. Recently, a new type of pluripotent stem cells was generated in vitro by Takahasi and Yamanaka [3]. By introducing only four transcription factors, *Oct3/4, Sox2, c-Myc, and Klf4*, differentiated cells could erase their cell identity, through utilization of Polycomb and Trithorax group complexes, thus producing the induced pluripotent stem cells (iPSCs). However, the use of *c-Myc* in this process, a known oncogene, may significantly hamper the clinical application of iPSCs in TE and RM strategies. With this in mind, more research is needed to be performed in this field in order for the development of iPSCs to become clinically safe. On the other hand, mesenchymal stromal cells (MSCs), a mesodermal population that can be derived from both adult and embryonic tissues, may be used as an alternative cellular population for TE approaches [4]. According to the International Society for Cellular Therapies (ISCT), MSCs are non-hematopoietic plastic adherent cells, which can be differentiated into "chondrocytes", "osteocytes", and "adipocytes" [5]. Immunophenotypically, MSCs are expressing CD73, CD90, and CD105, while lacking expression of CD34, CD45, and HLA-DR [5]. MSCs can be derived from various sources, including Wharton's jelly tissue, placental tissue, bone marrow, adipose tissue, dental pulp, liver, and lungs [4]. In addition, these cells can be easily expanded under in vitro culturing conditions for several passages without affecting their genome stability.

The field of TE relies on the use of various types of scaffolds, which can successfully mimic the biology of the extracellular matrix (ECM). Scaffolds provide a 3D microenvironment, where the cells can be adhered and proliferated under specific chemical and biophysical stimuli [6]. Furthermore, specialized bioreactor systems may contribute to the proper scaffold repopulation, cell proliferation and differentiation, even more. Scaffolds can be derived either from biological origin, including decellularized matrices, or can be fabricated in various dimensions, using mainly macromolecules derived from different origins, like expanded polytetrafluoroethylene (ePTFE), polyglycolic acid (PGA), polylactic acid (PLA), and polylactic co-glycolic acid (PLGA). The key properties of an ideal scaffold for TE can be summarized as (a) biocompatibility, (b) biodegradability, (c) mechanical properties, (d) easy fabrication, (e) non-toxic, and (f) proper cell attachment. Until now, scaffolds in combination with or without cells have been used in a wide variety of TE applications, including tendon and bone regeneration, blood vessel engineering, and trachea, heart, and esophagus development [7]. However, more research in this field must be performed by the scientific society in order to improve the clinical applications.

In this special issue of *Bioengineering*, we highlight the beneficial effects of stem cells and scaffolds in the emerging era of TE. The current issue included articles regarding the use of stem cells in TE and the proper production of biologically based scaffolds like nerve conduit, esophageal scaffold, and fibrin gel.

Under this scope, the immunoregulatory/immunosuppressive properties of MSCs that were derived from vitrified Wharton's Jelly tissue are shown (specifically, MSCs expressed successfully the HLA-G, a non-classical HLA class I molecule, which is considered to be the main immunosuppressive agent during pregnancy) [8]. In this way, the MSCs could be administrated in injured sites, reducing the host immune response, thus contributing to tissue regeneration. The beneficial regenerative properties of MSCs have also been described in the pilot study of Protogerou et al. [9]. In this study, MSCs in combination with platelet lysate were administrated to treat patients with erectile dysfunction (ED). The results showed the improvement of ED, which will be used for enrolling a wider study with a higher number of patients.

The use of iPSCs in TE and RM approaches might be very promising but still needs further clarification. Under this scope, Dias et al. [10] showed the dominant role of Wnt signaling in lineage commitment of human iPSCs. Moreover, the dominant effect of Wnt signaling over FGF and TGF-β was shown, resulting in the differentiation of iPSCs towards mesodermal lineages.

Regarding the biologically based scaffold development, Gontika et al. [11] described the utilization of decellularized human umbilical arteries (hUAs) as nerve conduits. Specifically, hUAs were obtained after gestation from umbilical cords and submitted to decellularization procedure. The produced scaffolds were free of cellular and nuclear material, while the ECM was preserved, as was observed by the histological analysis. Then, this scaffold was used as a nerve conduit in sciatic nerve injury. The results showed that the decellularized hUAs could support the elongation of nerve fibers and possibly could allow for the reinnervation of the target organs.

In a similar manner, the efficient development of a tissue engineered construct derived from rat esophagus was demonstrated [12]. In this study, rat esophagi were successfully decellularized. The ECM ultrastructure was retained after the decellularization procedure, and the obtained results could be used for scaling up this protocol to human tissues.

This special issue also included a preliminary study for fibrin gel production obtained from low volume cord blood units. The produced fibrin gel is characterized by several proteins that possibly contribute to tissue regeneration and possesses an alternative scaffold for wound healing.

Yuan et al. [13] introduce a new categorization method for scaffolds in order to avoid any misunderstandings between researchers. This new scaffold classification is of major importance, and is especially relevant in TE research.

The main scope of this special issue was to present state of the art tissue engineering approaches. Considerable effort has been undertaken by the scientific community toward the in vitro development of artificial organs such as heart, lungs, and liver [14]. The proper combination of stem cells and scaffolds under the conditions of good manufacturing practices (GMPs) could bring this form of personalized medicine one step closer to its clinical application. Finally, the editor would like to express his appreciation to the authors for their contribution to this special issue.

Conflicts of Interest: The authors declare no conflict of interest.

References

1. Langer, R.; Vacanti, J.P. Tissue Engineering. *Science* **1993**, *14*, 920–926. [CrossRef] [PubMed]
2. Daley, G.Q. Stem cells and the evolving notion of cellular identity. *Philos. Trans. R Soc. Lond B Biol. Sci.* **2015**, *19*, 370. [CrossRef]
3. Takahashi, K.; Yamanaka, S. Induction of pluripotent stem cells from mouse embryonic and adult fibroblast cultures by defined factors. *Cell* **2006**, *25*, 663–676. [CrossRef] [PubMed]
4. Chatzistamatiou, T.K.; Papassavas, A.C.; Michalopoulos, E.; Gamaloutsos, C.; Mallis, P.; Gontika, I.; Panagouli, E.; Koussoulakos, S.L.; Stavropoulos-Giokas, C. Optimizing isolation culture and freezing methods to preserve Wharton's jelly's mesenchymal stem cell (MSC) properties: An MSC banking protocol validation for the Hellenic Cord Blood Bank. *Transfusion* **2014**, *54*, 3108–3120. [CrossRef]
5. Dominici, M.; Le Blanc, K.; Mueller, I.; Slaper-Cortenbach, I.; Marini, F.C.; Krause, D.S.; Deans, R.J.; Keating, A.; Prockop, D.J.; Horwitz, E.M. Minimal criteria for defining multipotent mesenchymal stromal cells. *Cytotherapy* **2006**, *8*, 315–317. [CrossRef]
6. Salgado, A.J.; Oliveira, J.M.; Martins, A.; Teixeira, F.G.; Silva, N.A.; Neves, N.M.; Sousa, N.; Reis, R.L. Tissue engineering and regenerative medicine: Past, present, and future. *Int. Rev. Neurobiol.* **2013**, *108*, 1–33. [PubMed]
7. O'Brien, F.J. Biomaterials & scaffolds for tissue engineering. *Materialstoday* **2011**, *14*, 88–95.
8. Mallis, P.; Boulari, D.; Michalopoulos, E.; Dinou, A.; Spyropoulou-Vlachou, M.; Stavropoulos-Giokas, C. Evaluation of HLA-G Expression in Multipotent Mesenchymal Stromal Cells Derived from Vitrified Wharton's Jelly Tissue. *Bioengineering* **2018**, *5*, 95. [CrossRef] [PubMed]
9. Protogerou, V.; Michalopoulos, E.; Mallis, P.; Gontika, I.; Dimou, Z.; Liakouras, C.; Stavropoulos-Giokas, C.; Kostakopoulos, N.; Chrisofos, M.; Deliveliotis, C. Administration of Adipose Derived Mesenchymal Stem Cells and Platelet Lysate in Erectile Dysfunction: A Single Center Pilot Study. *Bioengineering* **2019**, *6*, 21. [CrossRef] [PubMed]
10. Dias, T.P.; Fernandes, T.G.; Diogo, M.M.; Cabral, J.M.S. Multifactorial Modeling Reveals a Dominant Role of Wnt Signaling in Lineage Commitment of Human Pluripotent Stem Cells. *Bioengineering* **2019**, *6*, 71. [CrossRef]
11. Gontika, I.; Katsimpoulas, M.; Antoniou, E.; Kostakis, A.; Stavropoulos-Giokas, C.; Michalopoulos, E. Decellularized Human Umbilical Artery Used as Nerve Conduit. *Bioengineering* **2018**, *5*, 100. [CrossRef] [PubMed]
12. Mallis, P.; Chachlaki, P.; Katsimpoulas, M.; Stavropoulos-Giokas, C.; Michalopoulos, E. Optimization of Decellularization Procedure in Rat Esophagus for Possible Development of a Tissue Engineered Construct. *Bioengineering* **2018**, *6*, 3. [CrossRef] [PubMed]
13. Yuan, H. Introducing the Language of "Relativity" for New Scaffold Categorization. *Bioengineering* **2019**, *6*, 20. [CrossRef] [PubMed]
14. Gilbert, T.W.; Sellaro, T.L.; Badylak, S.F. Decellularization of tissues and organs. *Biomaterials* **2006**, *27*, 3675–3683. [CrossRef] [PubMed]

© 2019 by the authors. Licensee MDPI, Basel, Switzerland. This article is an open access article distributed under the terms and conditions of the Creative Commons Attribution (CC BY) license (http://creativecommons.org/licenses/by/4.0/).

Article

Multifactorial Modeling Reveals a Dominant Role of Wnt Signaling in Lineage Commitment of Human Pluripotent Stem Cells

Tiago P. Dias [1,2], Tiago G. Fernandes [1,2], Maria Margarida Diogo [1,2] and Joaquim M. S. Cabral [1,2,*]

1 iBB—Institute for Bioengineering and Biosciences and Department of Bioengineering, Instituto Superior Técnico, Universidade de Lisboa, Av. Rovisco Pais, 1049-001 Lisbon, Portugal
2 The Discoveries Centre for Regenerative and Precision Medicine, Lisbon Campus, Instituto Superior Técnico, Universidade de Lisboa, Av. Rovisco Pais, 1049-001 Lisbon, Portugal
* Correspondence: joaquim.cabral@tecnico.ulisboa.pt

Received: 5 July 2019; Accepted: 13 August 2019; Published: 15 August 2019

Abstract: The human primed pluripotent state is maintained by a complex balance of several signaling pathways governing pluripotency maintenance and commitment. Here, we explore a multiparameter approach using a full factorial design and a simple well-defined culture system to assess individual and synergistic contributions of Wnt, FGF and TGFβ signaling to pluripotency and lineage specification of human induced pluripotent stem cells (hiPSC). Hierarchical clustering and quadratic models highlighted a dominant effect of Wnt signaling over FGF and TGFβ signaling, drawing hiPSCs towards mesendoderm lineages. In addition, a synergistic effect between Wnt signaling and FGF was observed to have a negative contribution to pluripotency maintenance and a positive contribution to ectoderm and mesoderm commitment. Furthermore, FGF and TGFβ signaling only contributed significantly for negative ectoderm scores, suggesting that the effect of both factors for pluripotency maintenance resides in a balance of inhibitory signals instead of proactive stimulation of hiPSC pluripotency. Overall, our dry-signaling multiparameter modeling approach can contribute to elucidate individual and synergistic inputs, providing an additional degree of comprehension of the complex regulatory mechanisms of human pluripotency and commitment.

Keywords: multiparameter; factorial design; Wnt signaling; TGFβ signaling; FGF signaling; human induced pluripotent stem cells; pluripotency and commitment

---

1. Introduction

Human induced pluripotent stem cells (hiPSCs) have an incredible potential for regenerative medicine therapies, drug-screening and disease modeling [1–3]. Understanding pluripotency and controlling commitment is essential to take full advantage of hiPSC properties and to develop efficient protocols to induce hiPSC direct differentiation into the cell types of interest.

Human pluripotency is usually associated with a primed state, controlled by a complex balance between multiple signaling pathways that govern pluripotency maintenance and exit from pluripotency towards differentiation [4–6]. This state has been connected with a weak stability and a bias towards commitment resembling the mouse epiblast state [7,8], contrasting with the increased stability of the naïve pluripotent state [9–11].

FGF, TGFβ and Wnt signaling pathways are among the most important pathways controlling hiPSC fate [4–6]. These signaling pathways can be associated with pleiotropic effects, stimulating divergent cellular responses such as self-renewal and commitment [12–14]. For example, the combined effects of FGF signaling and TGFβ signaling are typically associated with hiPSCs self-renewal [4,15]. Individually, however, FGF signaling has been connected with both neuroectoderm inhibition [16] and activation [7–9]. On the other hand, TGFβ signaling results in SMAD2/3 activation, which is associated with mesendoderm lineage specification [17,18]. Importantly, Wnt/β-catenin signaling is associated with self-renewal in hiPSCs [4,19–21], in line with being essential to promote the naïve pluripotency state and inhibit epiblast transition [10,11,22,23]. However, during differentiation, Wnt signaling is also associated to self-renewal disruption and guidance of cells towards mesendoderm commitment [24,25]. Also noteworthy is the fact that Wnt signaling has a role in directing cells from neuroectoderm towards neural crest specification [25,26], and that it inhibits cardiac mesoderm specification [27,28] while promoting the epicardial cell fate [29]. Furthermore, these signaling pathways can be interconnected and influenced by multiple signals at different pathway nodes, resulting in synergistic or antagonistic effects that can shift commitment towards specific lineages [30–33]. Thus, complex and undefined culture systems with multiple signaling inputs, often using conditioned media or serum, can provide a signaling overload, contributing to divergent and pleiotropic responses, that can mask the true impact of each signaling input. Development of a multiparameter approach with a controlled signaling environment can allow to fully discern the multiple singular and cooperative contributions of each signaling input allowing the identification of synergistic and antagonistic effects [34].

We previously used a multifactorial analysis approach that revealed a significant contribution of Wnt signaling to mESC pluripotency under physiological oxygen tensions [34]. Here, we use a dry-signaling multiparameter approach consisting of a full factorial design, combining the activation of Wnt, FGF and TGFβ signaling in hiPSCs cultured in a simple and well-defined culture system. Hierarchical clustering and quadratic models for human pluripotency and lineage commitment were designed and highlighted a Wnt signaling dominance with or without the presence of FGF and TGFβ inputs. Synergistic effects were observed between Wnt and FGF signaling by the pluripotency, ectoderm and mesoderm models. In addition, FGF and TGFβ signaling contributed negatively to the ectoderm model without a significant contribution for the pluripotency model, suggesting that a balanced inhibitory effect is promoting hiPSC pluripotency maintenance.

2. Materials and Methods

2.1. Human Induced Pluripotent Stem Cell Culture

In this work, the hiPSC cell line iPS-DF6-9-9T.B, purchased from WiCell Bank, was mainly used. This cell line is vector free and was derived from foreskin fibroblasts with a karyotype 46, XY. Both the hiPSC cell line F002.1A.13 provided by TCLab (Tecnologias Celulares para Aplicação Médica, Unipessoal, Lda.) that was generated using a retroviral system and the hiPSC line Gibco™ (Thermo Fisher Scientific, Waltham, MA, USA) derived from $CD34^+$ cells of healthy donors were used to validate results as described in the different sections and figure legends.

Maintenance of hiPSC culture was performed using an mTeSR1 medium (STEMCELL Technologies, Vancouver, BC, Canada) in 6-well tissue culture plates coated with Matrigel (BD Biosciences, San Jose, CA, USA) and diluted 1:30 in DMEM/F12. The medium was changed daily. Human iPSC passaging was performed using an EDTA (Thermo Fisher Scientific, Waltham, MA, USA) solution diluted in PBS at a concentration of 0.5 mM. Cells were incubated for 5 min with EDTA at room temperature and flushed with culture medium. For maintenance cultures, splits from 1:3 to 1:8 were usually performed. For cell counting, a sample of 100 μL was incubated in 400 μL of Accutase for 7 min at room temperature and samples were diluted 1:2 in Trypan Blue (Thermo Fisher Scientific, Waltham, MA, USA) for counting using a hemocytometer. Culture photos were obtained using a Leica DMI 3000B microscope (Leica Microsystems GmbH, Wetzlar, Germany) and a digital camera Nikon DXM 1200 (Nikon, Tokyo, Japan).

2.2. Full Factorial Design

A 3^3 full factorial design consisting of 27 culture conditions, corresponding to different concentrations of three different soluble factor activators of FGF, TGFβ and Wnt signaling (FGF2, TGFβ and CHIR, respectively), as well as three concentration levels (0, 1/3 and 1), was performed using E6 medium (Thermo Fisher Scientific, Waltham, MA, USA) as the basal medium. FGF2 (PeproTech, Rocky Hill, NJ, USA) concentration levels ranged from 0, 35 ng/mL to 100 ng/mL; TGFβ1 (PeproTech, Rocky Hill, NJ, USA) concentration levels ranged from 0, 0.7 ng/mL to 2 ng/mL; and CHIR99021 (Stemgent, Cambridge, MA, USA) concentration levels ranged from 0, 2 µM to 6 µM (Table 1). Three blocks of 9 culture conditions (samples) were performed each time with mTeSR1, E8 and E6 as controls. Cells were collected by EDTA Enzyme-free passaging and were seeded at 37,500 cells/cm^2 using an mTeSR1 medium, to guarantee that the results of the study would not be affected by cell confluence. Conditions were exposed to the respective cocktail after 24 h and fresh supplemented medium changed every 24 h for 4 consecutive days of exposure. Fresh medium was prepared every day and supplemented with the cytokines and small molecules prior to medium change. After 4 days of exposure, cells were singularized using Accutase for 7 min, centrifuged, and a sample counted to evaluate cell number fold increase (FI) using trypan blue. Cells were washed with PBS, centrifuged, and the cell pellets were stored at −80 °C to perform real-time PCR afterwards.

Table 1. Full factorial design conditions. FGF2 concentration levels ranged between 0, 35, and 100 ng/mL; TGFβ concentration levels ranged between 0, 0.85, and 2 ng/mL; and CHIR concentration levels ranged between 0, 2, and 6 µM.

Samples	FGF2 (ng/mL)	TGFβ (ng/mL)	CHIR (µM)
Sample 1/E6	0	0	0
Sample 2	0	0	2
Sample 3	0	0	6
Sample 4	0	0.7	0
Sample 5	0	0.7	2
Sample 6	0	0.7	6
Sample 7	0	2	0
Sample 8	0	2	2
Sample 9	0	2	6
Sample 10	35	0	0
Sample 11	35	0	2
Sample 12	35	0	6
Sample 13	35	0.7	0
Sample 14	35	0.7	2
Sample 15	35	0.7	6
Sample 16	35	2	0
Sample 17	35	2	2
Sample 18	35	2	6
Sample 19	100	0	0
Sample 20	100	0	2
Sample 21	100	0	6
Sample 22	100	0.7	0
Sample 23	100	0.7	2
Sample 24	100	0.7	6
Sample 25	100	2	0
Sample 26	100	2	2
Sample 27	100	2	6

2.3. Human iPSC-Cardiomyocyte (hiPSC-CM) Differentiation

Human iPSCs were seeded at a density of 1×10^5 cells/cm^2 and maintained in pluripotency conditions with daily medium changes. When confluence reached percentages around 95%, hiPSC cardiac differentiation was induced following the Wnt signaling modulation protocol previously described by Lian et al. [35]. Experiments were performed using 1 µM or 6 µM of the GSK3β inhibitor CHIR99021 (Stemgent, Cambridge, MA, USA) at day 0 and with or without 5 µM of the Wnt signaling inhibitor IWP4 (Stemgent, Cambridge, MA, USA) at day 3. Cells were collected and analyzed at day 15 of differentiation.

2.4. Human iPSC-Neural Differentiation

Human iPSCs were seeded at a density of 2×10^5 cells/cm^2 using E8. For E6 differentiation, after overnight growth, the medium was changed to E6 as previously described by Lippmann et al. [36]. For dual SMAD Inhibition-based neural induction, after cultures were nearly confluent, the medium was changed to 1:1 N2/B27 media supplemented with 10 µM SB431542 (Stemgent, Cambridge, MA, USA) and 100 nM LDN193189 (Stemgent, Cambridge, MA, USA), as previously described [37,38]. For both protocols, the medium was changed daily, and cells were collected and analyzed at day 12 of differentiation.

2.5. Flow Cytometry

Cells were washed with PBS, singularized and fixed using 2% (v/v) PFA for 20 min at room temperature. Cells were centrifuged and resuspended in 90% (v/v) cold methanol, incubated for 15 min at 4 °C. Samples were then washed 3 times using a solution of 0.5% (v/v) BSA in PBS (FB1). Primary antibody Cardiac Troponin T (CTNT) monoclonal mouse IgG antibody (Thermo Fisher Scientific, Waltham, MA, USA, Clone 13-11, dilution 1:250) or Primary antibody T/Brachyury polyclonal goat IgG antibody (R&D Systems, dilution 1:20) were diluted in FB1 plus 0.1% (v/v) Triton (FB2) and incubated for 1 h at room temperature. Cells were then washed and the cell pellet resuspended with the secondary antibody goat anti-mouse Alexa-488 (Thermo Fisher Scientific, Waltham, MA, USA) for CTNT or secondary antibody donkey anti-goat Alexa-488 for T/Brachyury (Thermo Fisher Scientific, Waltham, MA, USA), both diluted 1:1000 in FB2 and incubated for 30 min in the dark. Finally, cells were washed twice and cell pellets were resuspended in 500 µL of PBS and analyzed in a FACSCalibur flow cytometer (Becton Dickinson, Franklin Lakes, NJ, USA). Data were analyzed using the software "Flowing Software" at http://www.flowingsoftware.com (version 2.5).

2.6. Immunofluorescence Staining

Cells were fixed with 4% (v/v) PFA for 15 min, washed with PBS and incubated with blocking solution (10% v/v NGS, 0.1% v/v Triton-X in PBS) for 1 h. After incubation, for hiPSC-CM differentiation, Cardiac Troponin T (CTNT) monoclonal mouse IgG antibody (Thermo Fisher Scientific, Waltham, MA, USA, Clone 13-11) was diluted 1:250 in staining solution (5% v/v NGS, 0.1% v/v Triton-X in PBS) and incubated for 2 h at room temperature. For hiPSC-Neural commitment, NESTIN monoclonal mouse IgG antibody (R&D Systems, Minneapolis, MN, USA) and PAX6 polyclonal rabbit IgG antibody (Covance, Princeton, NJ, USA) were used both diluted 1:1000 in staining solution and incubated for 2 h at room temperature. After washing with PBS, secondary antibodies goat anti-mouse IgG Alexa-546 and goat anti-rabbit IgG Alexa-488 (Thermo Fisher Scientific, Waltham, MA, USA) were diluted 1:500 in staining solution and incubated for 1 h at room temperature. Samples were then washed 2 times with PBS, incubated for 2 min with 3 µg/mL of DAPI diluted in PBS, washed again 3 times, and stored at 4 °C. Samples were analyzed using a fluorescence optical microscope (Leica DMI 3000B, Leica Microsystems GmbH, Wetzlar, Germany) and a digital camera (Nikon DXM 1200, Nikon, Tokyo, Japan). Images were processed using ImageJ/Fiji (http://fiji.sc) [39] and PAX6$^+$ cells were quantified using CellProfiler (Broad Institute, Cambridge, MA, USA).

2.7. Real-Time PCR

RNA from each condition and controls was extracted using the High Pure RNA Isolation Kit (Roche, Basel, Switzerland) following the instructions provided with the Kit. RNA was quantified using a nanodrop, and 1 µg of RNA was converted to cDNA using the High Capacity cDNA Reverse Transcription Kit (Thermo Fisher Scientific, Waltham, MA, USA) following the instructions provided with the kit. Relative gene expression was evaluated using 10 ng of cDNA, 250 µM of each primer (Table S1) and using the Fast SYBR Green Master Mix (Thermo Fisher Scientific, Waltham, MA, USA) with an annealing temperature set to 60 °C. Melting curves were performed at the end to assess if primers were amplifying only the correct amplicon. Values were treated following the $2^{-\Delta\Delta CT}$ method.

GAPDH gene expression was used as endogenous control and relative expression was calibrated for each gene using mTeSR1 gene expression values. For hiPSC-CM differentiation, relative expression was calibrated using day 0 of differentiation.

For hiPSC-Neural commitment, real-time PCR was performed using the TaqMan Gene Expression Assay (Thermo Fisher Scientific, Waltham, MA, USA) for the genes *OCT4/POU5F1* (Hs00999634_gH), *NANOG* (Hs02387400_g1), *PAX6* (Hs00240871_m1), *SOX1* (Hs01057642_s1) and *GAPDH* (Hs02758991_g1). *GAPDH* gene expression was used as endogenous control and relative expression was calibrated using day 0 of differentiation.

2.8. Panels and Scores

Relative expression values were normalized using the minimum and maximum value obtained for each gene. Then, panels for pluripotency (*OCT4* and *NANOG*), ectoderm (*FGF5*, *PAX6* and *P75*), mesendoderm (*MIXL1* and *T*), mesoderm (*NKX2.5* and *MESP1*) and endoderm (*SOX17* and *PDX1*) were created by averaging the expression value of each gene. Then, scores for pluripotency and for each lineage were empirically calculated as follows:

$$\textbf{Pluripotency Score} = 1.5 \times \text{Pluripotent Panel} - 0.25 \times \text{Ectoderm Panel} \\ -0.25 \times \text{Mesendoderm Panel} - 0.5 \times \text{Mesoderm Panel} - 0.5 \times \text{Endoderm Panel}, \tag{1}$$

$$\textbf{Ectoderm Score} = 1.75 \times \text{Ectoderm Panel} - 0.25 \times \text{Pluripotent Panel} - 0.5 \times \\ \text{Mesendoderm Panel} - 0.5 \times \text{Mesoderm Panel} - 0.5 \times \text{Endoderm Panel}, \tag{2}$$

$$\textbf{Mesendoderm Score} = \text{Mesendoderm Panel} + 0.25 \times \text{Endoderm Panel} \\ + 0.25 \times \text{Mesoderm Panel} - 0.5 \times \text{Ectoderm Panel} - \text{Pluripotent Panel}, \tag{3}$$

$$\textbf{Endoderm Score} = 2 \times \text{Endoderm Panel} + 0.5 \times \text{Mesoderm Panel} - 0.5 \times \\ \text{Mesoderm Panel} - \text{Ectoderm Panel} - \text{Pluripotent Panel}, \tag{4}$$

$$\textbf{Mesoderm Score} = 2 \times \text{Mesoderm Panel} + 0.5 \times \text{Mesoderm Panel} - 0.5 \times \\ \text{Endoderm Panel} - \text{Ectoderm Panel} - \text{Pluripotent Panel}. \tag{5}$$

The main results showed in this study using scores were not changed when panels or individual gene expression were used. Nevertheless, scores helped to clarify the true effect of signal combinations, leading to more robust, statistically significant models.

2.9. Hierarchical Clustering and PCA

Hierarchical clusters and principal component analysis (PCA) were performed using Clustvis, a web tool based on R [40]. Clusters were obtained using Pearson correlation and average linkage. PCA were obtained using the Clustvis default SVD imputation.

2.10. Full Factorial Design Models and Statistical Analysis

A model for each score was created using Statistica Software. Models were obtained by fitting the data to a full quadratic model (linear, quadratic and two-way interactions) with centered and scaled polynomials, as follows:

$$Y_i = \beta_0 + \beta_1[X_1] + \beta_{11}[X_1]^2 + \beta_2[X_2] + \beta_{22}[X_2]^2 + \beta_3[X_3] + \beta_{33}[X_3]^2 + \beta_{12}[X_1][X_2] + \beta_{13}[X_1][X_3] + \beta_{23}[X_2][X_3] \tag{6}$$

where Y_i corresponds to the specific score; β_0 is the intersect coefficient; β_1, β_2 and β_3 are the coefficients correspondent to the linear main effects; β_{11}, β_{22} and β_{33} are the quadratic coefficients and β_{12}, β_{13} and β_{23} are the coefficients for factor interactions. The full factorial design with three replicates of Sample 1 (E6) resulted in a total of 28 degrees of freedom. Statistical significance for each model was assessed by ANOVA using Fisher's statistical test, in which factors with *p*-values lower than 0.05 were considered to have a statistically significant contribution to the model [34,41]. Models were not further refined by discarding non-statistically significant factors. Nevertheless, R^2-adjusted (R^2-Adj), a modified version

3. Results

3.1. Full Factorial Analysis in a "Dry-signaling" Culture System

To expose the impact of FGF signaling, TGF/Nodal signaling and Wnt signaling in human pluripotency and exit towards differentiation, a full factorial design was conceived to detect dual signaling roles by combining three concentration levels of each signaling input: Zero, lower activation (1/3 of higher activation) and higher activation, using E6/VTN [15], a dry-signaling system, as the basal culture medium (Figure 1). When compared with the E8 formulation [15], the E6 medium has only insulin as a principal signaling input, eliminating from its formulation FGF2 and Nodal/TGFβ. The experimental design covered 27 different conditions (Table 1). In addition, three replicates of each E6 basal media (Sample 1), mTeSR and E8 experiments were performed as controls.

In our multiparameter approach, FGF pathway was modulated using FGF2 at concentrations of 0, 35 and 100 ng/mL. Both TeSR and E8 medium use 100 ng/mL of FGF2 to maintain hiPSC pluripotency [15,42]. At this concentration and higher, a plateau of maximal activity is observed for downstream FGF signaling targets such as ERK and FRS-2 [12]. In fact, maximum activation of both downstream targets can be observed at 10 ng/mL, which can contribute to the pleiotropic behavior of FGF signaling [12]. In addition, TGF pathway was modulated using TGFβ1 at concentrations of 0, 0.7 and 2 ng/mL. E8 medium uses 1.74 ng/mL of TGFβ1 to maintain hiPSC pluripotency, although this concentration also has an impact in fibroblast proliferation and can inhibit hiPSC reprograming [43]. In TeSR, 0.6 ng/mL of TGFβ1 has a mild contribution to maintain pluripotency by directly targeting NANOG [17,42]. TGFβ1 at a concentration of 1 ng/mL seems to be enough to plateau maximum expression of downstream targets such as SMAD3 and release of IL-6 and CXCL8 [13]. Lastly, the Wnt pathway was modulated using the small chemical inhibitor CHIR99021 (CHIR) at concentrations of 0, 2 and 6 µM, which inhibits GSK3β leading to canonical Wnt signaling activation [44]. CHIR is one of the most potent and specific GSK3β inhibitors in vitro and seems to not significantly affect other kinases [44–47]. A concentration of 6 µM of CHIR is commonly used to promote hiPSC exit from pluripotency towards mesendoderm [28,35]. Lower concentrations, usually up to 2 µM, are found to be involved in self-renewal of human naïve PSCs [10], while 3 µM in the presence of dual SMAD inhibitors can induce hPSC neural crest differentiation [48].

Each condition of the full factorial design was assessed by analyzing the overall fold increase in total cell numbers, colony morphology and by real-time PCR, allowing the attribution of scores to each condition and the assessment of the data by clustering and modeling tools (Figure 1B). To assess the effect of each input, human iPSCs were seeded in VTN using mTeSR1. After 24 h, hiPSCs were exposed to the respective cocktail of signaling inputs using E6 as basal media. Exposure was performed for 4 days, changing the media every 24 h, to assess if specific cocktail combinations contributed to maintain pluripotency or guided cells to differentiation. Cell fold increase for all experimental conditions (samples) was evaluated after 5 days in culture (Figure 2A). All experimental conditions promoted cell growth, although, in general, cells exposed to media cocktails without CHIR presented a lower cell growth compared to conditions with CHIR supplementation. This result was consistent with the cell morphology observed, with cells exposed to cocktails containing CHIR showing a more differentiated-like phenotype at day 4 when compared with more well-defined compact colonies, typically associated with the pluripotent state, for cells without exposure to CHIR (Figure 2B). These cell morphology changes were observed gradually with increased CHIR exposure time (Figure S1). The effect of Wnt activation in colony morphology is in clear contrast with the effect of CHIR, at lower concentrations, commonly observed for mice or human cells in the naïve state of pluripotency with round-cells organized in a more compact and multilayer-like colony morphology [9,11,34,49].

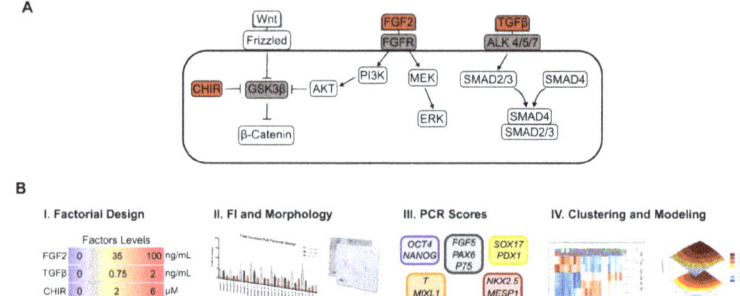

Figure 1. Schematic representation of the experimental framework used in this study. (**A**) A multiparameter approach was designed to reveal FGF, TGF/Nodal and Wnt signaling synergistic impact on human pluripotency and exit towards differentiation. (**B**) Multiparameter methodology performed. (i) A full factorial design combining 3 factors and 3 concentration levels for each factor was performed using E6 and Vitronectin as a dry-signaling culture system. Cells were exposed to the respective molecule cocktails for 4 days, with the medium changed daily. (ii) Proliferation and morphology were assessed for all 27 conditions plus mTeSR1 and E8 at the end of the 4-day culture. (iii) Real-time PCR was performed for each panel, and scores for pluripotency, ectoderm, mesendoderm, mesoderm and endoderm were calculated for each condition. Finally, (iv) scores obtained for each condition were hierarchical clustered and fitted to full quadratic models for each score.

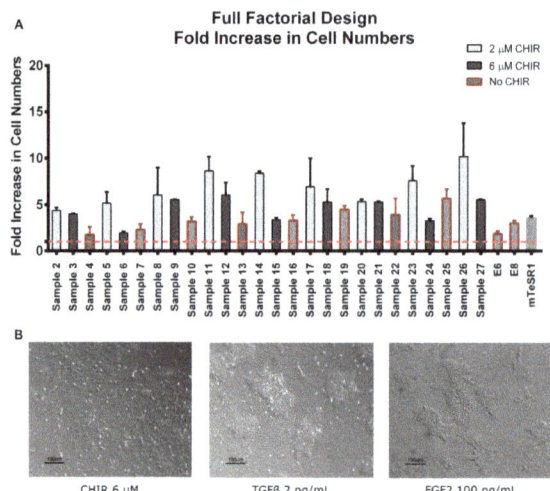

Figure 2. Human iPSC fold increase in total cell numbers and morphology of full factorial design conditions and controls. (**A**) Cell fold increase of all full factorial design conditions and controls. Red-dotted line marks the minimal threshold for fold increase achievement (FI = 1). In general, medium cocktails supplemented with CHIR showed higher cell fold increases when compared to cocktails without CHIR (highlighted in red). Error bars, standard error of the mean (SEM), $n = 2$. p-value < 0.01 by one-way ANOVA. (**B**) Typical morphology of cultures with CHIR addition compared to cultures without CHIR addition after 72 h of exposure to signaling inputs. Cultures without CHIR retained typical pluripotent colony morphology when compared to E8 or mTeSR1, while CHIR supplementation showed no colonies, which is associated with a more committed phenotype. See Table 1 for detailed concentrations of samples.

3.2. Pluripotency and Lineage Specification Evaluation Using Hierarchical Clustering and Principal Components Analysis

To assess the effect on cells exposed to each signaling input, real-time PCR was performed to analyze the expression of a set of genes (Table S1), corresponding to pluripotency and different lineage markers, whose expression levels could indicate if pluripotency was maintained or cells started to commit towards a specific lineage upon exposure to the different molecular cocktails. For the pluripotency panel, OCT4 and NANOG were selected, since both form the pluripotency core with SOX2, with OCT4 being enough to maintain and induce pluripotency [50] and NANOG being a sensible marker and gatekeeper of the pluripotent state [51–53]. Ectoderm panel was constituted by FGF5, a post-implantation primitive ectoderm marker [54]; PAX6, an early marker of neuroectodermal differentiation [55]; and P75, a neural crest cell marker [26]. Mesendoderm panel was composed by the primitive streak genes T/Brachyury, essential for primitive streak formation and mesendoderm differentiation, and MIXL1, a mesendoderm morphogen appearing at later stages of differentiation [56–58]. The endoderm panel was constituted by SOX17, a sensitive definitive endoderm marker [18,31], and PDX1, a foregut endoderm marker and regulator of pancreas specification [59]. The mesoderm panel was defined by MESP1, an early mesoderm marker that contributes to the specification of multiple mesoderm lineages in a context-dependent manner [60], and NKX2.5, a cardiac mesoderm marker expressed upon cardiac crescent formation [61].

To emphasize the main path that hiPSCs were following after exposure to the signaling cocktails, scores to each lineage commitment and pluripotency were attributed to each sample. This data was hierarchically clustered using Pearson correlation and average linkage, resulting in two main clusters mainly explained by the presence or absence of Wnt signaling activation (Figure 3A). Cocktails exposing hiPSCs to CHIR clustered together and led to higher mesendoderm, endoderm and mesoderm scores, while conditions without CHIR clustered together and led to higher pluripotency and ectoderm scores. In addition, PCAs show that 91.3% of data variability (PC1) rely on Wnt signaling variation (Figure 3B). The exception was hiPSCs exposed to mTeSR1, which has LiCl (0.1 mM), a Wnt activator [42], and registered higher pluripotency and ectoderm scores, with lower mesendoderm, mesoderm and endoderm scores. These results seem to further highlight colony morphology observations: Wnt signaling activation was imposing hiPSCs to exit pluripotency and to commit towards mesendoderm lineages with or without FGF and TGF activation.

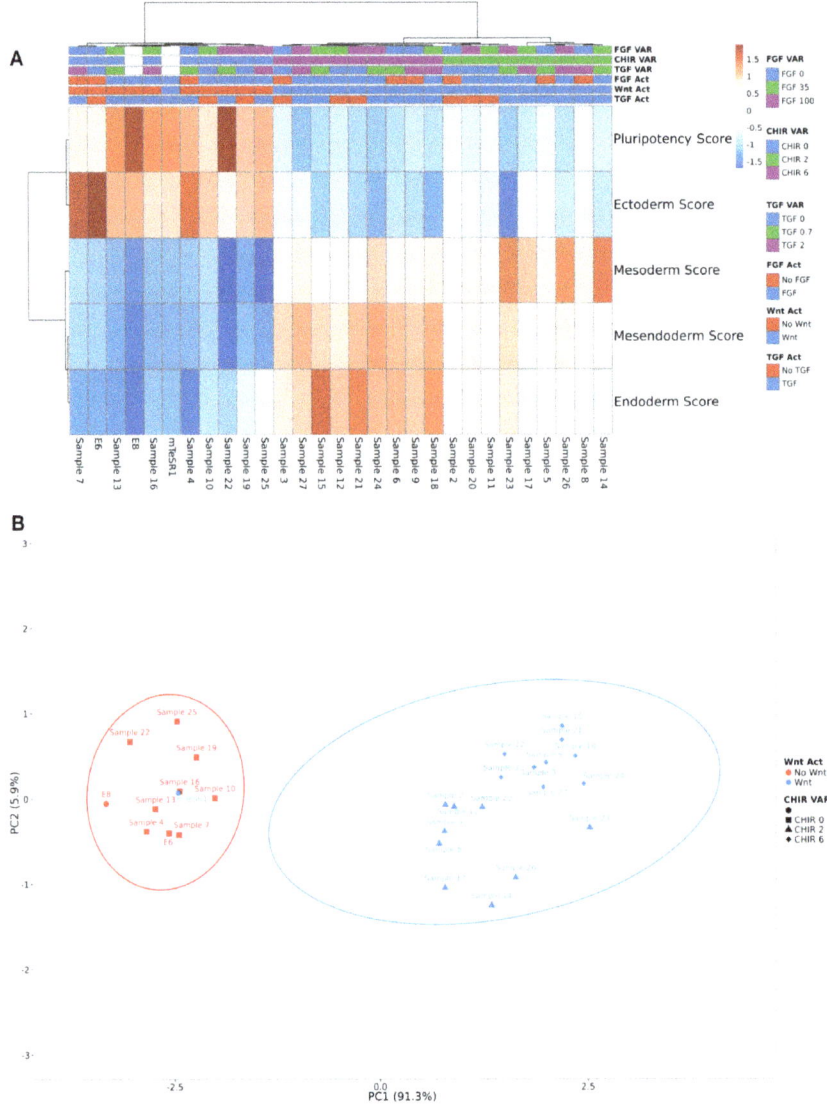

Figure 3. Hierarchical clustering and PCAs of full factorial design scores revealed two main clusters concordant with the presence or absence of CHIR. (**A**) Two main clusters were observed: Wnt activation, characterized by higher scores to mesendoderm lineages; and No Wnt activation, characterized by higher pluripotency and ectoderm scores. Clustering was performed using Pearson correlation and average linkage. Samples were labelled using signaling activation and no activation (FGF Act, Wnt Act and TGF Act; red and blue) and concentration variation (FGF VAR, CHIR VAR and TGF VAR; blue, green and purple). (**B**) The two main principal components together explained 97.2% of total data set variance. PC1 variance (91.3% of data set) corresponds to Wnt modulation. Only mTeSR1 clustered within the "No Wnt" group.

3.3. Full Quadratic Models for the Pluripotency and Ectoderm Lineage Scores

For visualizing the true impact of Wnt signaling in each score and discern if any synergies between signaling pathways were present, full quadratic models were fitted to the data, including quadratic, linear and two-way interactions. Contribution of each factor was considered statistically significant for p-values < 0.05. As expected, CHIR supplementation contributed negatively for pluripotency scores (Figure 4A–C). Additionally, another significant contribution highlighted by the model could be observed, with synergy of FGF2 and CHIR contributing to lower pluripotency scores. Similar negative effects of CHIR supplementation were observed for ectoderm scores (Figure 4D–F). Furthermore, the FGF2 linear term and the TGFβ quadratic term also contributed negatively to ectoderm scores. This result is in line with FGF2 showing to repress *PAX6* [16] and TGF inhibition facilitating neuroectoderm differentiation [55]. Contrarily, a synergy of FGF2 and CHIR contributed to higher ectoderm scores, which is coherent with reports showing that this synergy can lead to ectodermal neural crest and placode lineages [30].

To further explore the differences between ectoderm induction and pluripotency maintenance highlighted in our models, hiPSC were differentiated using a combination of dual SMAD inhibitors [55], ensuring inhibition of BMP and TGFβ autocrine stimulation, and compared with cells differentiated in E6 medium only, therefore allowing hiPSCs to follow their inner circuitry and autocrine path [36]. Cells with no inhibitors (E6) showed similar profiles compared with neural differentiation induced with inhibitors (Dual), although showing a slight delay in the decrease of pluripotent markers *OCT4* and *NANOG*, and in the increase of *SOX1* (Figure 4G). This was reflected in PAX6$^+$ cells originated at day 12 for all three cell lines tested, with dual SMAD inhibition resulting in a 20% to 40% increase in neural progenitors (Figure 4H,I). Nevertheless, cell differentiated in E6 medium only originated significant amounts of PAX6$^+$ cells as well (Figure 4H). In fact, multiple neural rosettes were observed at day 12 for all the three hiPSC cell lines differentiated in E6 (Figure 4J), suggesting that this condition can allow a high degree of neural progenitor organization and commitment (Figure 4I) [36]. These results show the natural tendency for hiPSCs to converge to ectoderm if not actively stimulated [36,62], and are in line with the signaling inputs contributing significantly for both models.

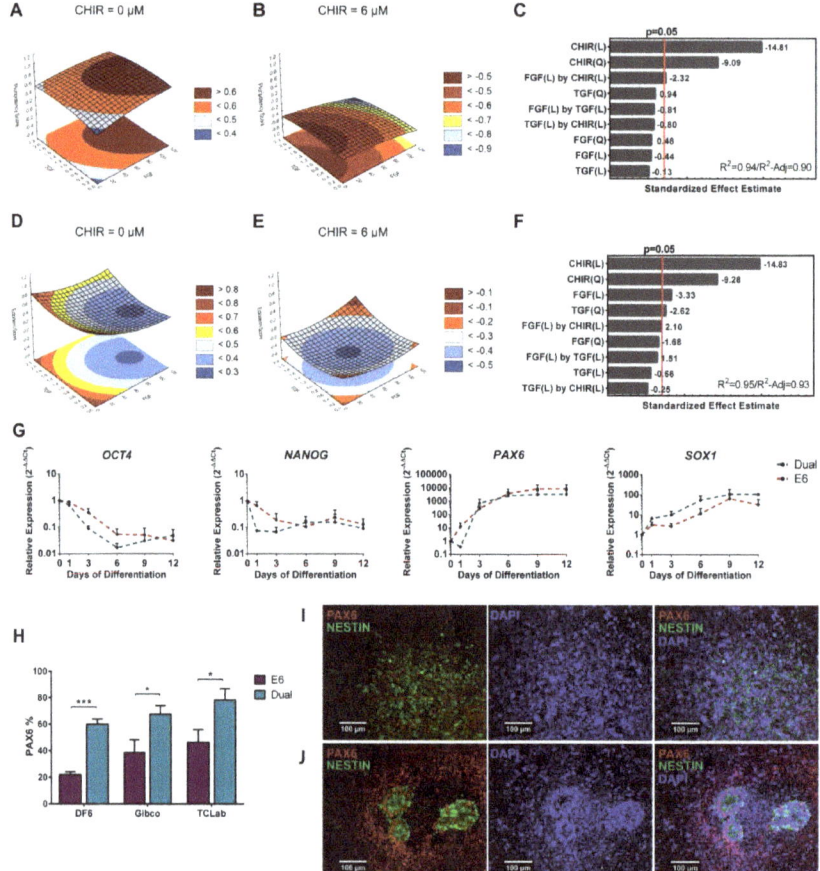

Figure 4. Quadratic models for the pluripotency and ectoderm scores highlighted a dominant negative contribution of Wnt signaling. (**A,B**) Representative curves of TGFβ and FGF2 contributions to pluripotency model with CHIR set at zero (**A**) and at 6 µM (**B**). Without CHIR, FGF2 high concentrations resulted in higher scores in the model, while with CHIR set at 6 µM, both TGFβ and FGF2 presence decreases pluripotency score. (**C**) CHIR linear and quadratic terms are the ones that contributed the most to the model, decreasing pluripotency scores. A statistically significant negative synergy can be seen between CHIR and FGF2. Model showed a good fit with a R^2 of 0.94 and a R^2-Adjusted of 0.90. (**D,E**) Representative curves of TGFβ and FGF2 contributions to ectoderm model with CHIR set to zero (**D**) and to 6 µM (**E**). Without CHIR, the model output higher ectoderm scores, concordant with the significant negative effect. (**F**) Besides CHIR linear and quadratic negative effects, FGF2 linear, TGFβ quadratic and an interaction between CHIR and FGF2 contributed significantly to the model. FGF2 positively contributed when conjugated with CHIR, while negatively contributed for ectoderm score without CHIR. TGFβ negative quadratic term contribution can be clearly observed when FGF2 is zero (**D**) with higher ectoderm scores at full or no activation. Model showed a good fit to the data set with a R^2 of 0.95 and a R^2-Adjusted of 0.93. (**G**) Real-time PCR comparison of E6 and dual SMAD inhibition neuroectoderm differentiation revealed a delayed expression decrease of the pluripotency markers *OCT4* and *NANOG* for the differentiation without factors (E6). Error bars, SEM, $n = 3$. (**H**) PAX6 positive cells quantification showed that dual SMAD differentiation originated 20% to 40% more PAX6$^+$ cells for the three cell lines tested, when compared with E6 differentiation at day 12. Error bars, SEM, $n = 7$ DF6 and Gibco, and $n = 5$ TCLab. * p-value < 0.05, *** p-value < 0.001 (two-sided t-test). (**I,J**) At day 12, dual SMAD (**I**) differentiation did not result in neural rosette formation while cells differentiated using only E6 (**J**) consistently organized in neural rosettes throughout the culture. Scale bar: 100 µm. See Figures S2 and S3, and Tables S2 and S3 for pluripotency and ectoderm full model information.

3.4. Full Quadratic Models for the Mesendoderm and Mesoderm Lineage Scores

For the mesendoderm score model (Figure 5A–C), an inverse contribution from CHIR linear and quadratic terms was observed when compared with the pluripotency and ectoderm models, concordant with the results showed by the hierarchical clustering and PCAs. CHIR terms were the only components of the model that were statistically significant (Figure 5C). Input of 1/3 (2 µM) of CHIR registered a steep increase in mesendoderm scores with the full input doubling the score (Figure 5D). To further understand and validate the mesendoderm model, cardiac differentiation was performed using the full level of CHIR (6 µM) and compared with a low level of Wnt Activation (1 µM), since hiPSCs died in the differentiation conditions without CHIR. Only 6 µM of CHIR contributed significantly for the expression of the mesendoderm transcription factor T/Brachyury with a peak at day 1 of both protein (Figure 5E) and mRNA (Figure 5F), despite both conditions contributing similarly to a decrease in *OCT4* (Figure 5F). The observation that 1 µM of CHIR is insufficient for the entrance into mesendoderm is in full concordance with the model obtained, which predicts negative mesendoderm scores for that level of Wnt activation with or without the activation of TGF and FGF signaling (Figure S4).

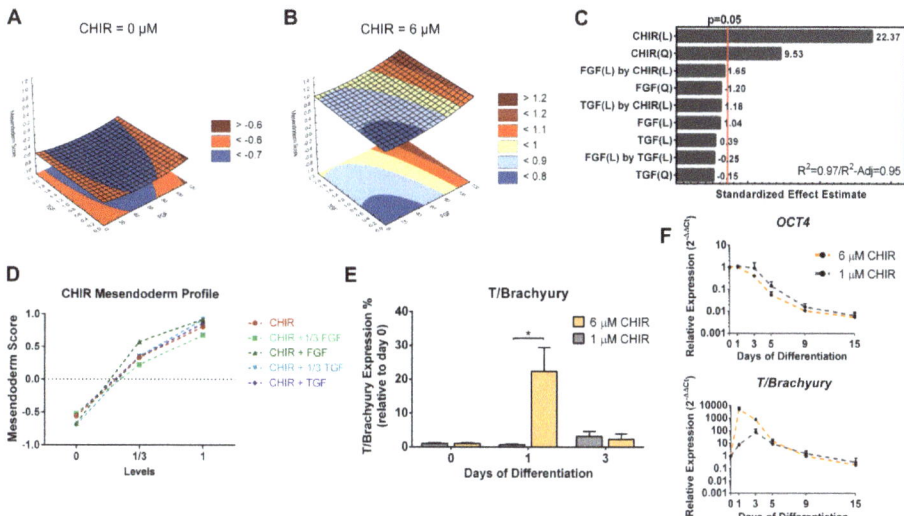

Figure 5. Quadratic model for the mesendoderm scores highlighted a strong positive contribution of Wnt signaling. (**A,B**) Representative curves of TGFβ and FGF2 contributions to the mesendoderm model with CHIR set to zero (**A**) and to 6 µM (**B**). (**C**) CHIR linear and quadratic terms of the model significantly contributed to the mesendoderm model. The model showed a very good fit to the data set, with a R^2 of 0.97 and a R^2-Adjusted of 0.95. (**D**) Mesendoderm score profile of CHIR supplemented conditions shows an increase with CHIR concentration. (**E**) Flow cytometry of T/Brachyury showed that expression is significantly higher when 6 µM of CHIR is used compared to a lower activation level (1 µM). Error bars, SEM, n = 3. * p-value < 0.05 (two-sided t-test). (**F**) Real-time PCR comparison of *OCT4* and T/Brachyury for cardiac differentiation using 6 µM or 1 µM of CHIR showed a similar decreasing profile of *OCT4* gene expression, while 6 µM of CHIR contributed to a significantly higher gene expression of T/Brachyury at day 1. Error bars, SEM, n = 3. See Figure S4 and Table S4 for mesendoderm full model information.

Similarly, CHIR linear and quadratic terms also contributed significantly for the mesoderm score model (Figure 6A–C). In addition, FGF and CHIR linear terms present a positive synergy in the cutoff of statistical significance for the model (p = 0.06), which can be observed in Figure 6B.

This synergy is in agreement with reports showing that dual activation of FGF and Wnt promotes hiPSC differentiation into mesenchymal stem cells [63] and the generation of neuromesodermal progenitors [64,65]. FGF signaling also had a positive contribution for endoderm scores, with the FGF linear term and both CHIR quadratic and linear terms having a significant positive impact in the scores (Figures S6 and S7). This is in line with reports showing that both factors are essential to efficient definitive endoderm differentiation [31], and with FGF signaling pathway playing an important role in further differentiating the definitive endoderm, particularly into liver, lung and pancreatic lineages [18,59,66].

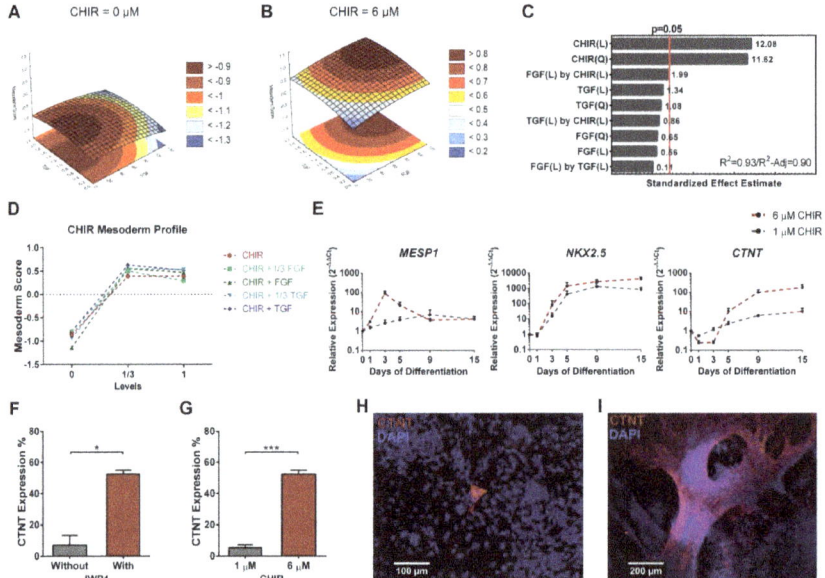

Figure 6. Quadratic model for the mesoderm scores highlighted the contribution of Wnt signaling with higher scores for intermediate CHIR concentrations. (**A,B**) Representative curves of TGFβ and FGF2 contributions to the mesoderm model with CHIR set at zero (**A**) and at 6 µM (**B**). (**C**) CHIR quadratic and linear terms significantly contributed to the mesoderm model. A synergy of CHIR with FGF can also be observed. Model showed a good fit to the data set with a R^2 of 0.93 and a R^2-Adjusted of 0.90. (**D**) CHIR mesoderm profile shows an increase in mesoderm score at 1/3 activation, while higher concentrations maintain or slightly decrease mesoderm scores. (**E**) Real-time PCR comparison of cardiac differentiation using 6 µM or 1 µM of CHIR registered a higher gene expression of *MESP1*, with a peak at day 3, *NKX2.5* and *CTNT* when 6 µM is used. Error bars, SEM, $n = 3$. (**F**) Flow cytometry of cardiac differentiation with or without IWP4 showed that inhibiting Wnt signaling at day 3 is essential to efficiently obtain CTNT positive cells. Error bars, SEM, $n = 3$. * p-value < 0.05 (two-sided t-test). (**G**) Flow cytometry comparing Wnt signaling low or high activation levels showed that initial low activation originated few CTNT positive cells compared to 6 µM of CHIR. Error bars, SEM, $n = 3$. *** p-value < 0.001 (two-sided t-test). (**H,I**) Consistent with flow cytometry, immunostaining showed that 1 µM of CHIR (H, scale bar 100 µm) originated few cardiomyocytes while 6 µM originated cardiomyocytes throughout the culture (I, scale bar 200 µm). See Figure S5 and Table S5 for mesoderm full model information.

Contrarily to the mesendoderm model, stimulation with 2 µM of CHIR gave rise to the higher mesoderm scores, while the full input of 6 µM disclosed a tendency to stagnate or even decrease such scores (Figure 6D). For the generation of cardiac mesoderm, stimulation with 6 µM of CHIR resulted in increased *MESP1* expression, with a peak at day 3. Furthermore, *NKX2.5* expression was not significantly affected by the level of Wnt signaling stimulation, but *CTNT* expression was one order of magnitude higher at 6 µM when compared with 1 µM of CHIR (Figure 6E). Concomitant with our model prediction, later inhibition of Wnt signaling, using IWP4 [27], was fundamental to obtain hiPSC-derived cardiomyocytes (Figure 6F), an observation that can be predicted in our model by the maintenance or even decrease in mesoderm scores when a full level input of CHIR was used (Figure 6D and Figure S5). Further aligned with the model predictions, low activation of Wnt signaling originated a low number of cardiomyocytes (Figure 6G), particularly sparse and rarely observable (Figure 6H) when compared to the full level of input (Figure 6I).

4. Discussion

There is a multitude of signaling pathways and interactions that govern pluripotency maintenance and lineage specification. Uncontrolled and poorly defined systems with increased noise to signal ratio hinder the ability to fully understand the independent and synergistic role of each factor in hPSC fate. The focus of our study was to develop a multiparameter approach to study the individual and synergistic effect of Wnt, FGF and TGFβ signaling pathways using a dry-signaling culture system to avoid major unspecific signaling contributions. Our results showed that Wnt signaling had a dominant effect over FGF and TGFβ inputs, pulling hiPSCs away from pluripotency and ectoderm, towards mesendoderm lineages. In addition, a synergy between FGF and Wnt signaling was observed, with a negative contribution to pluripotency scores and a positive contribution to ectoderm and mesoderm scores. FGF and TGFβ signaling negatively contributed to ectoderm scores, which is connected with the well-known role of these signaling pathways in maintaining hiPSCs epiblast-like pluripotent state [4,15], and preventing cells to follow their inner circuitry towards neuroectoderm [36,62] (Figure 7).

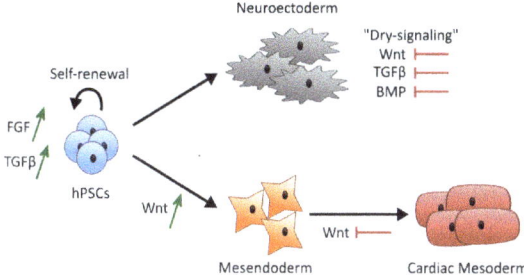

Figure 7. Model summarizing the overall results obtained using a dry-signaling multiparameter approach. Wnt signaling activation showed to be dominant over FGF and TGFβ signaling driving hPSCs towards mesendoderm lineages. A synergy of FGF and CHIR was observed providing higher ectoderm scores or higher mesoderm scores and contributing to lower pluripotency scores. Contribution of FGF and TGFβ signaling to maintain pluripotency scores seems to be connected with the negative contribution of both FGF and TGFβ signaling to ectoderm scores, with absence of inputs inducing cells to follow their inner circuitry towards neuroectoderm.

In our model, the linear contribution of FGF and the quadratic term of TGFβ show a negative correlation to the ectoderm score, with the lowest result obtained for levels of 100 ng/mL of FGF (full input) and 0.7 ng/mL of TGFβ (1/3 of input). These values are in concordance with the level of input provided by both commercial media E8 and TeSR used to maintain hiPSCs pluripotency [15,42]. In addition, these levels of input prevent hiPSCs to naturally exit pluripotency towards neuroectoderm as observed by us and previously reported by Lippmann and coworkers [36], and are in concordance with reports showing that both factors inhibit neuroectoderm differentiation [16,55]. As suggested from our results, the role of FGF and TGFβ in pluripotency maintenance seems to derive from a thin balance that prevents exit towards differentiation, instead of actively promoting and stimulating pluripotency (Figure 4). This seems to be in line with the weak stability of the pluripotent epiblast-like state [7,8], and the bias of hPSCs towards neuroectoderm when only both factors are present to maintain pluripotency [36,62].

In the presence of Wnt signaling stimulation, TGF and FGF signaling effects were secondary, with all models showing CHIR terms as the most significant contributors. This dominance contributed negatively for pluripotency and ectoderm scores (Figure 4), and positively for mesendoderm, mesoderm and endoderm scores (Figures 5 and 6, Figures S5 and S6). These results are coherent with the literature and the well-known importance of Wnt signaling in the commitment of hPSCs towards mesendoderm and lineage specifications that rely on Wnt activation, such as cardiac mesoderm [27], pancreatic β-cells [67], mesenchymal stem cells [63], or epicardial lineage cells [29]. Besides contributing to commitment, Wnt signaling is described to have a role in promoting self-renewal and the naïve state of pluripotency [9–11,34]. Our pluripotency model predicted higher pluripotency scores for input levels lower than 2 µM, but was unable to fully register a positive contribution of Wnt signaling to pluripotency. This might be explained by the epiblast-like state of the hiPSCs used in our study, and the inability of Wnt activation to reprogram cells to the naïve state by itself [10,11,49]. Once in a primed state, Wnt signaling role seems to transition from self-renewal to promoting further commitment to mesendoderm [23].

Synergies observed between Wnt signaling and FGF signaling in our models are also coherent with previously reported data. The positive contribution for the ectoderm model is in line with the role of FGF in repressing *PAX6* [16] and, together with Wnt, synergistically promoting the specification towards ectodermal neural crest and placode lineages [26,30]. The proximity between the pluripotent state with neuroectoderm specification can explain the negative synergistic impact of both pathways in the pluripotency model. In addition, contribution to lower pluripotency scores is also in line with the impact of Wnt in promoting mesendoderm lineages, and FGF being important to specify mesendoderm towards endoderm lineages, which is also coherent with the predictions of our endoderm model [18,31,59,66,68]. Lastly, the synergistic effect of both pathways in the mesoderm model is in line with the paraxial specification and direct differentiation of hiPSCs towards mesenchymal and neuromesodermal progenitors [32,63–65].

In conclusion, using a multifactorial, multiparameter modeling approach we predicted a dominant role of Wnt signaling over FGF and TGF signaling in our dry-signaling culture system. This modeling methodology also allowed the construction of models providing a rational understanding of hiPSCs pluripotency and commitment, allowing to discriminate the different synergies between FGF and Wnt signaling, in agreement with previously reported studies. Following this proposed framework, carefully designed 5-level fractional factorial designs coupled with multiple signaling activation dynamics should contribute to the construction of models with increased sensitivity and reduced variance and, consequently, providing an extra degree of comprehension of the complex regulatory system of human pluripotency and commitment.

Supplementary Materials: The following are available online at http://www.mdpi.com/2306-5354/6/3/71/s1, Table S1: Primer pairs used for real-time PCR, Table S2: ANOVA Pluripotency score model, Table S3: ANOVA Ectoderm score model, Table S4: ANOVA Mesendoderm score model, Table S5: ANOVA Mesoderm score model, Table S6: ANOVA Endoderm score model, Figure S1: Cell morphology changes during cocktail exposure, Figure S2: Full panel of the quadratic model for the pluripotency scores highlighting a dominant negative contribution of Wnt signaling, Figure S3: Full panel of the quadratic model for the ectoderm scores highlighting a dominant negative contribution of Wnt signaling with FGF signaling also contributing to lower ectoderm scores, Figure S4: Full panel of the quadratic model for the mesendoderm scores highlighting a strong and dominant contribution of Wnt signaling, Figure S5: Full panel of the quadratic model for the mesoderm scores highlighting the contribution of Wnt signaling with higher scores for intermediate CHIR concentrations, Figure S6: Full panel of the quadratic model for the endoderm scores highlighted the contribution of Wnt signaling with FGF signaling also positively contributing to higher endoderm scores, Figure S7: Endoderm Model Score profiles and Standardized Effect Estimate.

Author Contributions: Conceptualization, T.P.D., T.G.F., M.M.D. and J.M.S.C.; methodology, T.P.D.; investigation, T.P.D.; formal analysis, T.P.D.; visualization, T.P.D.; writing—original draft preparation, T.P.D.; writing—review and editing, T.G.F., M.M.D. and J.M.S.C.; supervision, T.G.F., M.M.D. and J.M.S.C.

Funding: Funding was received by iBB through Programa Operacional Regional de Lisboa 2020 (Project N. 007317), through the EU COMPETE Programme and from National Funds through FCT under the Programme grant (SAICTPAC/0019/2015, MITP-TB/ECE/0013/2013, UID/BIO/04565/2013 and PTDC/EMD-TLM/29728/2017), by the DISCOVERIES CTR from EU Teaming Phase2 (H2020-WIDESPREAD-01-2016-2017).

Acknowledgments: Tiago P. Dias acknowledge Fundação para a Ciência e a Tecnologia (FCT, Portugal, http://www.fct.pt) for financial support (SFRH/BD/78774/2011).

Conflicts of Interest: The authors declare no conflict of interest.

References

1. Engle, S.J.; Puppala, D. Integrating human pluripotent stem cells into drug development. *Cell Stem Cell* **2013**, *12*, 669–677. [CrossRef]
2. Merkle, F.T.; Eggan, K. Modeling human disease with pluripotent stem cells: From genome association to function. *Cell Stem Cell* **2013**, *12*, 656–668. [CrossRef] [PubMed]
3. Fernandes, T.G.; Rodrigues, C.A.V.; Diogo, M.M.; Cabral, J.M.S. Stem cell bioprocessing for regenerative medicine. *J. Chem. Technol. Biotechnol.* **2014**, *89*, 34–47. [CrossRef]
4. Ludwig, T.E.; Levenstein, M.E.; Jones, J.M.; Berggren, W.T.; Mitchen, E.R.; Frane, J.L.; Crandall, L.J.; Daigh, C.A.; Conard, K.R.; Piekarczyk, M.S.; et al. Derivation of human embryonic stem cells in defined conditions. *Nat. Biotechnol.* **2006**, *24*, 185–187. [CrossRef] [PubMed]
5. Vallier, L.; Alexander, M.; Pedersen, R.A. Activin/Nodal and FGF pathways cooperate to maintain pluripotency of human embryonic stem cells. *J. Cell Sci.* **2005**, *118*, 4495–4509. [CrossRef] [PubMed]
6. Wang, L.; Schulz, T.C.; Sherrer, E.S.; Dauphin, D.S.; Shin, S.; Nelson, A.M.; Ware, C.B.; Zhan, M.; Song, C.-Z.; Chen, X.; et al. Self-renewal of human embryonic stem cells requires insulin-like growth factor-1 receptor and ERBB2 receptor signaling. *Blood* **2007**, *110*, 4111–4119. [CrossRef] [PubMed]
7. Kunath, T.; Saba-El-Leil, M.K.; Almousailleakh, M.; Wray, J.; Meloche, S.; Smith, A. FGF stimulation of the Erk1/2 signalling cascade triggers transition of pluripotent embryonic stem cells from self-renewal to lineage commitment. *Development* **2007**, *134*, 2895–2902. [CrossRef]
8. Tesar, P.J.; Chenoweth, J.G.; Brook, F.A.; Davies, T.J.; Evans, E.P.; Mack, D.L.; Gardner, R.L.; McKay, R.D. New cell lines from mouse epiblast share defining features with human embryonic stem cells. *Nature* **2007**, *448*, 196–199. [CrossRef]
9. Ying, Q.-L.; Wray, J.; Nichols, J.; Batlle-Morera, L.; Doble, B.; Woodgett, J.; Cohen, P.; Smith, A. The ground state of embryonic stem cell self-renewal. *Nature* **2008**, *453*, 519–523. [CrossRef]
10. Ware, C.B.; Nelson, A.M.; Mecham, B.; Hesson, J.; Zhou, W.; Jonlin, E.C.; Jimenez-Caliani, A.J.; Deng, X.; Cavanaugh, C.; Cook, S.; et al. Derivation of naive human embryonic stem cells. *Proc. Natl. Acad. Sci. USA* **2014**, *111*, 4484–4489. [CrossRef]
11. Gafni, O.; Weinberger, L.; Mansour, A.A.; Manor, Y.S.; Chomsky, E.; Ben-Yosef, D.; Kalma, Y.; Viukov, S.; Maza, I.; Zviran, A.; et al. Derivation of novel human ground state naive pluripotent stem cells. *Nature* **2013**, *504*, 282–286. [CrossRef] [PubMed]
12. Garcia-Maya, M.; Anderson, A.A.; Kendal, C.E.; Kenny, A.V.; Edwards-Ingram, L.C.; Holladay, A.; Saffell, J.L. Ligand concentration is a driver of divergent signaling and pleiotropic cellular responses to FGF. *J. Cell. Physiol.* **2006**, *206*, 386–393. [CrossRef] [PubMed]
13. O'Leary, L.; Sevinç, K.; Papazoglou, I.M.; Tildy, B.; Detillieux, K.; Halayko, A.J.; Chung, K.F.; Perry, M.M. Airway smooth muscle inflammation is regulated by microRNA-145 in COPD. *Febs Lett.* **2016**, *590*, 1324–1334. [CrossRef] [PubMed]
14. Blauwkamp, T.A.; Nigam, S.; Ardehali, R.; Weissman, I.L.; Nusse, R. Endogenous Wnt signalling in human embryonic stem cells generates an equilibrium of distinct lineage-specified progenitors. *Nat. Commun.* **2012**, *3*, 1070. [CrossRef] [PubMed]
15. Chen, G.; Gulbranson, D.R.; Hou, Z.; Bolin, J.M.; Ruotti, V.; Probasco, M.D.; Smuga-Otto, K.; Howden, S.E.; Diol, N.R.; Propson, N.E.; et al. Chemically defined conditions for human iPSC derivation and culture. *Nat. Methods* **2011**, *8*, 424–429. [CrossRef] [PubMed]

16. Greber, B.; Coulon, P.; Zhang, M.; Moritz, S.; Frank, S.; Müller-Molina, A.J.; Araúzo-Bravo, M.J.; Han, D.W.; Pape, H.-C.; Schöler, H.R. FGF signalling inhibits neural induction in human embryonic stem cells. *EMBO J.* **2011**, *30*, 4874–4884. [CrossRef] [PubMed]
17. Xu, R.-H.; Sampsell-Barron, T.L.; Gu, F.; Root, S.; Peck, R.M.; Pan, G.; Yu, J.; Antosiewicz-Bourget, J.; Tian, S.; Stewart, R.; et al. NANOG Is a Direct Target of TGFβ/Activin-Mediated SMAD Signaling in Human ESCs. *Cell Stem Cell* **2008**, *3*, 196–206. [CrossRef] [PubMed]
18. Teo, A.K.K.; Ali, Y.; Wong, K.Y.; Chipperfield, H.; Sadasivam, A.; Poobalan, Y.; Tan, E.K.; Wang, S.T.; Abraham, S.; Tsuneyoshi, N.; et al. Activin and BMP4 synergistically promote formation of definitive endoderm in human embryonic stem cells. *Stem Cells* **2012**, *30*, 631–642. [CrossRef] [PubMed]
19. Sato, N.; Meijer, L.; Skaltsounis, L.; Greengard, P.; Brivanlou, A.H. Maintenance of pluripotency in human and mouse embryonic stem cells through activation of Wnt signaling by a pharmacological GSK-3-specific inhibitor. *Nat. Med.* **2004**, *10*, 55–63. [CrossRef] [PubMed]
20. Ding, V.M.Y.; Ling, L.; Natarajan, S.; Yap, M.G.S.; Cool, S.M.; Choo, A.B.H. FGF-2 modulates Wnt signaling in undifferentiated hESC and iPS cells through activated PI3-K/GSK3beta signaling. *J. Cell. Physiol.* **2010**, *225*, 417–428. [CrossRef] [PubMed]
21. Cai, L.; Ye, Z.; Zhou, B.Y.; Mali, P.; Zhou, C.; Cheng, L. Promoting human embryonic stem cell renewal or differentiation by modulating Wnt signal and culture conditions. *Cell Res.* **2007**, *17*, 62–72. [CrossRef] [PubMed]
22. ten Berge, D.; Kurek, D.; Blauwkamp, T.; Koole, W.; Maas, A.; Eroglu, E.; Siu, R.K.; Nusse, R. Embryonic stem cells require Wnt proteins to prevent differentiation to epiblast stem cells. *Nat. Cell Biol.* **2011**, *13*, 1070–1075. [CrossRef] [PubMed]
23. Xu, Z.; Robitaille, A.M.; Berndt, J.D.; Davidson, K.C.; Fischer, K.A.; Mathieu, J.; Potter, J.C.; Ruohola-Baker, H.; Moon, R.T. Wnt/β-catenin signaling promotes self-renewal and inhibits the primed state transition in naïve human embryonic stem cells. *Proc. Natl. Acad. Sci. USA* **2016**, *113*, E6382–E6390. [CrossRef] [PubMed]
24. Davidson, K.C.; Adams, A.M.; Goodson, J.M.; McDonald, C.E.; Potter, J.C.; Berndt, J.D.; Biechele, T.L.; Taylor, R.J.; Moon, R.T. Wnt/β-catenin signaling promotes differentiation, not self-renewal, of human embryonic stem cells and is repressed by Oct4. *Proc. Natl. Acad. Sci. USA* **2012**, *109*, 4485–4490. [CrossRef] [PubMed]
25. Funa, N.S.; Schachter, K.A.; Lerdrup, M.; Ekberg, J.; Hess, K.; Dietrich, N.; Honoré, C.; Hansen, K.; Semb, H. β-Catenin Regulates Primitive Streak Induction through Collaborative Interactions with SMAD2/SMAD3 and OCT4. *Cell Stem Cell* **2015**, *16*, 639–652. [CrossRef] [PubMed]
26. Menendez, L.; Yatskievych, T.A.; Antin, P.B.; Dalton, S. Wnt signaling and a Smad pathway blockade direct the differentiation of human pluripotent stem cells to multipotent neural crest cells. *Proc. Natl. Acad. Sci. USA* **2011**, *108*, 19240–19245. [CrossRef] [PubMed]
27. Lian, X.; Hsiao, C.; Wilson, G.; Zhu, K.; Hazeltine, L.B.; Azarin, S.M.; Raval, K.K.; Zhang, J.; Kamp, T.J.; Palecek, S.P. Robust cardiomyocyte differentiation from human pluripotent stem cells via temporal modulation of canonical Wnt signaling. *Proc. Natl. Acad. Sci. USA* **2012**, *109*, E1848–E1857. [CrossRef] [PubMed]
28. Dias, T.P.; Pinto, S.N.; Santos, J.I.; Fernandes, T.G.; Fernandes, F.; Diogo, M.M.; Prieto, M.; Cabral, J.M.S. Biophysical study of human induced Pluripotent Stem Cell-Derived cardiomyocyte structural maturation during long-term culture. *Biochem. Biophys. Res. Commun.* **2018**, *499*, 611–617. [CrossRef] [PubMed]
29. Bao, X.; Lian, X.; Qian, T.; Bhute, V.J.; Han, T.; Palecek, S.P. Directed differentiation and long-term maintenance of epicardial cells derived from human pluripotent stem cells under fully defined conditions. *Nat. Protoc.* **2017**, *12*, 1890–1900. [CrossRef] [PubMed]
30. Tchieu, J.; Zimmer, B.; Fattahi, F.; Amin, S.; Zeltner, N.; Chen, S.; Studer, L. A Modular Platform for Differentiation of Human PSCs into All Major Ectodermal Lineages. *Cell Stem Cell* **2017**, *21*, 399–410. [CrossRef]
31. Loh, K.M.; Ang, L.T.; Zhang, J.; Kumar, V.; Ang, J.; Auyeong, J.Q.; Lee, K.L.; Choo, S.H.; Lim, C.Y.Y.; Nichane, M.; et al. Efficient Endoderm Induction from Human Pluripotent Stem Cells by Logically Directing Signals Controlling Lineage Bifurcations. *Cell Stem Cell* **2014**, *14*, 237–252. [CrossRef] [PubMed]
32. Loh, K.M.M.; Chen, A.; Koh, P.W.W.; Deng, T.Z.Z.; Sinha, R.; Tsai, J.M.M.; Barkal, A.A.A.; Shen, K.Y.Y.; Jain, R.; Morganti, R.M.M.; et al. Mapping the Pairwise Choices Leading from Pluripotency to Human Bone, Heart, and Other Mesoderm Cell Types. *Cell* **2016**, *166*, 451–468. [CrossRef] [PubMed]

33. Singh, A.M.; Reynolds, D.; Cliff, T.; Ohtsuka, S.; Mattheyses, A.L.; Sun, Y.; Menendez, L.; Kulik, M.; Dalton, S. Signaling network crosstalk in human pluripotent cells: A Smad2/3-regulated switch that controls the balance between self-renewal and differentiation. *Cell Stem Cell* **2012**, *10*, 312–326. [CrossRef] [PubMed]
34. Barbosa, H.S.C.; Fernandes, T.G.; Dias, T.P.; Diogo, M.M.; Cabral, J.M.S. New Insights into the Mechanisms of Embryonic Stem Cell Self-Renewal under Hypoxia: A Multifactorial Analysis Approach. *PLoS ONE* **2012**, *7*, e38963. [CrossRef] [PubMed]
35. Lian, X.; Zhang, J.; Azarin, S.M.; Zhu, K.; Hazeltine, L.B.; Bao, X.; Hsiao, C.; Kamp, T.J.; Palecek, S.P. Directed cardiomyocyte differentiation from human pluripotent stem cells by modulating Wnt/β-catenin signaling under fully defined conditions. *Nat. Protoc.* **2013**, *8*, 162–175. [CrossRef] [PubMed]
36. Lippmann, E.S.; Estevez-Silva, M.C.; Ashton, R.S. Defined human pluripotent stem cell culture enables highly efficient neuroepithelium derivation without small molecule inhibitors. *Stem Cells* **2013**, *32*, 1–18. [CrossRef] [PubMed]
37. Shi, Y.; Kirwan, P.; Livesey, F.J. Directed differentiation of human pluripotent stem cells to cerebral cortex neurons and neural networks. *Nat. Protoc.* **2012**, *7*, 1836–1846. [CrossRef]
38. Fernandes, T.G.; Duarte, S.T.; Ghazvini, M.; Gaspar, C.; Santos, D.C.; Porteira, A.R.; Rodrigues, G.M.C.; Haupt, S.; Rombo, D.M.; Armstrong, J.; et al. Neural commitment of human pluripotent stem cells under defined conditions recapitulates neural development and generates patient-specific neural cells. *Biotechnol. J.* **2015**, *10*, 1578–1588. [CrossRef]
39. Schindelin, J.; Arganda-Carreras, I.; Frise, E.; Kaynig, V.; Longair, M.; Pietzsch, T.; Preibisch, S.; Rueden, C.; Saalfeld, S.; Schmid, B.; et al. Fiji: An open-source platform for biological-image analysis. *Nat. Methods* **2012**, *9*, 676–682. [CrossRef]
40. Metsalu, T.; Vilo, J. ClustVis: a web tool for visualizing clustering of multivariate data using Principal Component Analysis and heatmap. *Nucleic Acids Res.* **2015**, *43*, W566–W570. [CrossRef]
41. Box, G.E.P.; Hunter, J.S.; Hunter, W.G. *Statistics for Experimenters: Design, Innovation, and Discovery*, 2nd ed.; Wiley-Interscience: Hoboken, NJ, USA, 2005; ISBN 978-0-471-71813-0.
42. Ludwig, T.E.; Bergendahl, V.; Levenstein, M.E.; Yu, J.; Probasco, M.D.; Thomson, J.A. Feeder-independent culture of human embryonic stem cells. *Nat. Methods* **2006**, *3*, 637–646. [CrossRef] [PubMed]
43. Beers, J.; Gulbranson, D.R.; George, N.; Siniscalchi, L.I.; Jones, J.; Thomson, J.A.; Chen, G. Passaging and colony expansion of human pluripotent stem cells by enzyme-free dissociation in chemically defined culture conditions. *Nat. Protoc.* **2012**, *7*, 2029–2040. [CrossRef] [PubMed]
44. Ring, D.B.; Johnson, K.W.; Henriksen, E.J.; Nuss, J.M.; Goff, D.; Kinnick, T.R.; Ma, S.T.; Reeder, J.W.; Samuels, I.; Slabiak, T.; et al. Selective Glycogen Synthase Kinase 3 Inhibitors Potentiate Insulin Activation of Glucose Transport and Utilization In Vitro and In Vivo. *Diabetes* **2003**, *52*, 588–595. [CrossRef] [PubMed]
45. Cline, G.W.; Johnson, K.; Regittnig, W.; Perret, P.; Tozzo, E.; Xiao, L.; Damico, C.; Shulman, G.I. Effects of a novel glycogen synthase kinase-3 inhibitor on insulin-stimulated glucose metabolism in Zucker diabetic fatty (fa/fa) rats. *Diabetes* **2002**, *51*, 2903–2910. [CrossRef]
46. Cohen, P.; Goedert, M. GSK3 inhibitors: Development and therapeutic potential. *Nat. Rev. Drug Discov.* **2004**, *3*, 479–487. [CrossRef] [PubMed]
47. Bain, J.; Plater, L.; Elliott, M.; Shpiro, N.; Hastie, C.J.; Mclauchlan, H.; Klevernic, I.; Arthur, J.S.C.; Alessi, D.R.; Cohen, P. The selectivity of protein kinase inhibitors: A further update. *Biochem. J.* **2007**, *408*, 297–315. [CrossRef] [PubMed]
48. Chambers, S.M.; Mica, Y.; Lee, G.; Studer, L.; Tomishima, M.J. Dual-SMAD Inhibition/WNT Activation-Based Methods to Induce Neural Crest and Derivatives from Human Pluripotent Stem Cells. In *Human Embryonic Stem Cell Protocols. Methods in Molecular Biology*; Turksen, K., Ed.; Humana Press: New York, NY, USA, 2013; pp. 329–343.
49. Hanna, J.; Cheng, A.W.; Saha, K.; Kim, J.; Lengner, C.J.; Soldner, F.; Cassady, J.P.; Muffat, J.; Carey, B.W.; Jaenisch, R. Human embryonic stem cells with biological and epigenetic characteristics similar to those of mouse ESCs. *Proc. Natl. Acad. Sci. USA* **2010**, *107*, 9222–9227. [CrossRef] [PubMed]
50. Hammachi, F.; Morrison, G.M.; Sharov, A.A.; Livigni, A.; Narayan, S.; Papapetrou, E.P.; O'Malley, J.; Kaji, K.; Ko, M.S.H.; Ptashne, M.; et al. Transcriptional activation by Oct4 is sufficient for the maintenance and induction of pluripotency. *Cell Rep.* **2012**, *1*, 99–109. [CrossRef]
51. Nichols, J.; Smith, A. Naive and primed pluripotent states. *Cell Stem Cell* **2009**, *4*, 487–492. [CrossRef]

52. Chambers, I.; Colby, D.; Robertson, M.; Nichols, J.; Lee, S.; Tweedie, S.; Smith, A. Functional expression cloning of Nanog, a pluripotency sustaining factor in embryonic stem cells. *Cell* **2003**, *113*, 643–655. [CrossRef]
53. Silva, J.; Smith, A. Capturing pluripotency. *Cell* **2008**, *132*, 532–536. [CrossRef] [PubMed]
54. Noisa, P.; Ramasamy, T.S.; Lamont, F.R.; Yu, J.S.L.; Sheldon, M.J.; Russell, A.; Jin, X.; Cui, W. Identification and Characterisation of the Early Differentiating Cells in Neural Differentiation of Human Embryonic Stem Cells. *PLoS ONE* **2012**, *7*, e37129. [CrossRef] [PubMed]
55. Chambers, S.M.; Fasano, C.A.; Papapetrou, E.P.; Tomishima, M.; Sadelain, M.; Studer, L. Highly efficient neural conversion of human ES and iPS cells by dual inhibition of SMAD signaling. *Nat. Biotechnol.* **2009**, *27*, 275–280. [CrossRef] [PubMed]
56. Wang, L.; Chen, Y.-G. Signaling Control of Differentiation of Embryonic Stem Cells toward Mesendoderm. *J. Mol. Biol.* **2016**, *428*, 1409–1422. [CrossRef] [PubMed]
57. Xi, Q.; Wang, Z.; Zaromytidou, A.-I.; Zhang, X.H.-F.; Chow-Tsang, L.-F.; Liu, J.X.; Kim, H.; Barlas, A.; Manova-Todorova, K.; Kaartinen, V.; et al. A poised chromatin platform for TGF-β access to master regulators. *Cell* **2011**, *147*, 1511–1524. [CrossRef] [PubMed]
58. Hart, A.H.; Hartley, L.; Sourris, K.; Stadler, E.S.; Li, R.; Stanley, E.G.; Tam, P.P.L.; Elefanty, A.G.; Robb, L. Mixl1 is required for axial mesendoderm morphogenesis and patterning in the murine embryo. *Development* **2002**, *129*, 3597–3608.
59. Johannesson, M.; Ståhlberg, A.; Ameri, J.; Sand, F.W.; Norrman, K.; Semb, H. FGF4 and retinoic acid direct differentiation of hESCs into PDX1-expressing foregut endoderm in a time- and concentration-dependent manner. *PLoS ONE* **2009**, *4*, e4794. [CrossRef]
60. Chan, S.S.-K.; Shi, X.; Toyama, A.; Arpke, R.W.; Dandapat, A.; Iacovino, M.; Kang, J.; Le, G.; Hagen, H.R.; Garry, D.J.; et al. Mesp1 Patterns Mesoderm into Cardiac, Hematopoietic, or Skeletal Myogenic Progenitors in a Context-Dependent Manner. *Cell Stem Cell* **2013**, *12*, 587–601. [CrossRef]
61. Den Hartogh, S.C.; Schreurs, C.; Monshouwer-Kloots, J.J.; Davis, R.P.; Elliott, D.A.; Mummery, C.L.; Passier, R. Dual Reporter MESP1 mCherry/w -NKX2-5 eGFP/w hESCs Enable Studying Early Human Cardiac Differentiation. *Stem Cells* **2015**, *33*, 56–67. [CrossRef]
62. Nazareth, E.J.P.; Ostblom, J.E.E.; Lücker, P.B.; Shukla, S.; Alvarez, M.M.; Oh, S.K.W.; Yin, T.; Zandstra, P.W. High-throughput fingerprinting of human pluripotent stem cell fate responses and lineage bias. *Nat. Methods* **2013**, *10*, 1225–1231. [CrossRef]
63. Winston, T.S.; Suddhapas, K.; Wang, C.; Ramos, R.; Soman, P.; Ma, Z. Serum-Free Manufacturing of Mesenchymal Stem Cell Tissue Rings Using Human-Induced Pluripotent Stem Cells. *Stem Cells Int.* **2019**, *2019*, 1–11. [CrossRef] [PubMed]
64. Turner, D.A.; Hayward, P.C.; Baillie-Johnson, P.; Rue, P.; Broome, R.; Faunes, F.; Martinez Arias, A. Wnt/β-catenin and FGF signalling direct the specification and maintenance of a neuromesodermal axial progenitor in ensembles of mouse embryonic stem cells. *Development* **2014**, *141*, 4243–4253. [CrossRef] [PubMed]
65. Gouti, M.; Tsakiridis, A.; Wymeersch, F.J.; Huang, Y.; Kleinjung, J.; Wilson, V.; Briscoe, J. In Vitro Generation of Neuromesodermal Progenitors Reveals Distinct Roles for Wnt Signalling in the Specification of Spinal Cord and Paraxial Mesoderm Identity. *Plos Biol.* **2014**, *12*, e1001937. [CrossRef] [PubMed]
66. Serls, A.E.; Doherty, S.; Parvatiyar, P.; Wells, J.M.; Deutsch, G.H. Different thresholds of fibroblast growth factors pattern the ventral foregut into liver and lung. *Development* **2005**, *132*, 35–47. [CrossRef] [PubMed]
67. Kunisada, Y.; Tsubooka-Yamazoe, N.; Shoji, M.; Hosoya, M. Small molecules induce efficient differentiation into insulin-producing cells from human induced pluripotent stem cells. *Stem Cell Res.* **2012**, *8*, 274–284. [CrossRef] [PubMed]
68. Hannan, N.R.F.; Segeritz, C.-P.; Touboul, T.; Vallier, L. Production of hepatocyte-like cells from human pluripotent stem cells. *Nat. Protoc.* **2013**, *8*, 430–437. [CrossRef] [PubMed]

© 2019 by the authors. Licensee MDPI, Basel, Switzerland. This article is an open access article distributed under the terms and conditions of the Creative Commons Attribution (CC BY) license (http://creativecommons.org/licenses/by/4.0/).

Technical Note

Short Term Results of Fibrin Gel Obtained from Cord Blood Units: A Preliminary in Vitro Study

Panagiotis Mallis [1,†], Ioanna Gontika [1,†], Zetta Dimou [1], Effrosyni Panagouli [1], Jerome Zoidakis [2], Manousos Makridakis [2], Antonia Vlahou [2], Eleni Georgiou [1], Vasiliki Gkioka [1], Catherine Stavropoulos-Giokas [1] and Efstathios Michalopoulos [1,*]

1. Hellenic Cord Blood Bank, Biomedical Research Foundation Academy of Athens, 4 Soranou Ephessiou Street, 115 27 Athens, Greece
2. Biotechnology Division, Biomedical Research Foundation Academy of Athens, 4 Soranou Ephessiou Street, 115 27 Athens, Greece
* Correspondence: smichal@bioacademy.gr; Tel.: +30-210-6597331; Fax: +30-210-6597345
† These authors contributed equally to this work as first authors.

Received: 20 June 2019; Accepted: 31 July 2019; Published: 2 August 2019

Abstract: Background: Recent findings have shown that the fibrin gel derived from cord blood units (CBUs) play a significant role in wound healing and tissue regeneration. The aim of this study was to standardize the fibrin gel production process in order to allow for its regular use. Methods: CBUs ($n = 200$) were assigned to 4 groups according to their initial volume. Then, a two-stage centrifugation protocol was applied in order to obtain platelet rich plasma (PRP). The concentration of platelets (PLTs), white blood cells (WBCs) and red blood cells (RBCs) were determined prior to and after the production process. In addition, targeted proteomic analysis using multiple reaction monitoring was performed. Finally, an appropriate volume of calcium gluconate was used in PRP for the production of fibrin gel. Results: The results of this study showed that high volume CBUs were characterized by greater recovery rates, concentration and number of PLTs compared to the low volume CBUs. Proteomic analysis revealed the presence of key proteins for regenerative medicine. Fibrin gel was successfully produced from CBUs of all groups. Conclusion: In this study, low volume CBUs could be an alternative source for the production of fibrin gel, which can be used in multiple regenerative medicine approaches.

Keywords: fibrin gel; platelet rich plasma; cord blood units; platelets; TGF-β1; proteomic analysis

---

1. Introduction

Fibrin gel, a platelet rich plasma (PRP) derivative, is mostly obtained from adult peripheral blood (APB) after a single blood collection [1–3]. Recently, cord blood has been proposed as an alternative source for the production of fibrin gel [4–6]. Fibrin gel constitutes a natural biomaterial that can be applied in various regenerative medicine approaches, including skin, cartilage and bone regeneration, wound healing and drug delivery [7–9]. This biomaterial can be fabricated with different degrees of stiffness, viscosity and degradation rate through the concentration adjustment of key elements such as calcium gluconate, sodium chloride (NaCl) and batroxobin [1] Recently, it has been shown that fibrin gel's stiffness was dependent on divalent cations of Ca and Mg, and on NaCl concentration [1].

The beneficial properties of fibrin gel in regenerative medicine are mostly attributed to the increased release of several growth factors—including transforming growth factor-β1 (TGF-β1), fibroblast growth factor (FGF), platelet derived growth factor A and B (PDGF AB and BB), vascular endothelial growth factor-A (VEGF-A) and hepatocyte growth factor (HGF)—through activation of platelets (PLTs) [9–11]. Specifically, these growth factors are contained in α-granules of PLTs, which are secreted during the blood coagulation cascade [12]. In this context, the entrapped PLTs in the platelet plug at the site of injury are activated by the FGF and PDGF produced from the injured cells [7,12,13]. The content of α-granules are then released, thus contributing even more in wound healing [7,12,13].

Fibrin formation is initiated when thrombin, an enzyme that is abundant in plasma, cleaves fibrinogen [7,14–16]. Fibrinogen (Fb) is a 340 kDa plasma glycoprotein which consists of three pairs of polypeptide chains $(A\alpha B\beta\gamma)_2$, and two types of fibrinopeptides A (FpA) and B (FpB), connected with 29 disulphide bonds to the Fb central region [7,14–16]. During the clotting cascade, thrombin cleaves the FpA and FpB between Arg–Gly residues [14–16]. This in turn exposes the knobs A and B of the central region, enabling the connection of fibrin monomers [7,14–16]. Further association of fibrin monomers results in the development of a three-dimensional network know as fibrin. This fibrin gel is a viscoelastic scaffold, and its properties are dependent on clot structure, number of entrapped PLTs and concentration of released growth factors [14–16].

Fibrin gel can be produced efficiently from cord blood units that do not meet the acceptable criteria of cord blood banks (CBBs) for processing and cryostorage [6,11,17]. It is estimated that over 80% of the received cord blood units (CBUs) are characterized by low number of total nucleated cells (TNCs <120×10^7 cells) and CD34 cells (<4×10^7 cells), low volume (<120 mL) and low viability (<85%), thus cannot be further processed [11,18]. These discarded CBUs could serve as a source for fibrin gel production under good manufacturing practices (GMP) conditions. Furthermore, CBUs contain the same amount of PLTs (150–400 × 10^3 PLTs/μL) as the adult peripheral blood (APB) samples, and PLTs from both have been characterized by the same proteomic profile [7,11,17]. Only some minor changes have been reported in PLTs derived from CBUs, regarding LYN, MAP3K5, and FAM129A signaling pathways, but without having any significant impact on their function [19].

A number of CBBs have focused on the production of fibrin gel, in order to use it as a tool for regenerative medicine approaches, worldwide [6,11,17]. However, the fibrin gel production varied between different research groups, and their results are still under debate [3,4,6]. Proper PRP isolation and fibrin gel production can be performed from CBUs with volume up to 82 mL. This is mostly attributed by the limitations of the PLTs isolation procedure. Specifically, the fibrin gel must contain high PLT number, while the red blood cell (RBC) content must be low [3,4,6]. In low volume CBUs, the PLTs cannot be isolated without having no RBC contamination, resulting to the rejection of this product. Under those circumstances, the CBUs that are used from the CBBs for fibrin gel production are restricted only to those with volume up to 82 mL and a low content of TNCs. In addition, high volume CBUs are characterized by a greater number of PLTs, thus could be the most valuable source for fibrin gel production. Although, high volume CBUs are characterized by great number of TNCs, and are more valuable being used as hematopoietic stem cell transplants, rather than as a source for fibrin gel development [6]. Based on the above data, and knowing that more than 80% of received CBUs are characterized by low volume and TNC number, it would be worthwhile to use them as a source for the fibrin gel production. For this purpose, the aim of this study was the standardization of the procedure for fibrin gel development, utilizing the low volume CBUs.

2. Methods

2.1. Cord Blood Units Collection

CBUs were collected at full-term gestation (37–40 weeks), after parental informed consent, in a collection bag (Macopharma, United Kingdom) containing 30 mL of citrate-phosphate-dextrose (CPD), and immediately distributed to Hellenic Cord Blood Bank (HCBB). The collections were performed in accordance with the ethical standards of the Greek National Ethical Committee, were approved by our Institution's ethical board (Referene No. 1508, 5/9/2018) and were in accordance with the declaration of Helsinki. After storage at 4 °C, the units were processed within 24 h from collection. Only the CBUs that did not meet the criteria outlined by HCBB (Table S1), were used for the preparation of the PRP.

2.2. Preparation of Platelet Rich Plasma and Fibrin Gel

In this study, n = 200, CBUs were distributed into four groups (n = 40 for each group) according to their initial net volume (including the anticoagulant). Group A, <81 mL; group B, 82–110 mL; group C, 111–148 mL; and group D involved CBUs that were pooled according to their blood group at a final volume of 111–148 mL. Specifically, in group D, 2 compatible CBUs were combined in order to produce 1 pooled CBU. The total number of CBUs that were combined in group D was 80, resulting in 40 pooled CBUs.

Automated cell counting was performed in all CBUs of each group with a hematological analyzer (Abacus 5 part 380 series, Beckman Coulter, Atlanta, GA, USA) for the determination of the total number and concentration of white blood cells (WBCs), RBCs and PLTs. CBUs with a PLT concentration of less than $150 \times 10^3/\mu L$, were excluded from the study [6].

Then, the CBUs were centrifuged at 210 g for 15 min, at room temperature (RT), and the top plasma fraction was collected and transferred in a secondary processing bar. A sample from the plasma fraction was taken for PLT counting at the hematological analyzer (Abacus 5 part 380, Beckman Coulter, Atlanta, GA, USA). Then, the plasma fraction was centrifuged at 2600 g for 15 min, at RT, and the supernatant platelet-poor plasma (PPP) was collected, in excess of the final target volume of the PRP. For the determination of the final volume of the PRP, the following equations were used, as previously described, with some modifications [6]:

$$\text{Lower limit of PRP volume} = \text{PLT concentration in plasma fraction}/(800 \times 10^3/\mu L)$$

$$\text{Upper limit of PRP volume} = \text{PLT concentration in plasma fraction}/(1400 \times 10^3/\mu L)$$

$$\text{Target PRP volume (mL)} = (\text{Upper limit} + \text{lower limit})/2$$

where, the factors $(800 \times 10^3/\mu L)$ and $(1400 \times 10^3/\mu L)$ corresponded to the minimum and maximum PLT concentration in PRP products. Specifically, the PLT concentration in the final product should be more than 800×10^3 PLTs/μL and not exceed 1400×10^3 PLTs/μL, otherwise the PRP units must be discarded.

Finally, the fibrin gel was obtained after the activation of PRP samples with the addition of 10% calcium gluconate, in ratio 3:1 at 37 °C, as has been described in literature [7,17].

2.3. Protein Determination Using Multiple Reaction Monitoring

In order to verify if significant alterations were presented in growth factor and chemokine content between PRP derived from CBUs and APB, liquid chromatography/multiple reaction monitoring (LC/MRM) analysis was applied. For this purpose, PRP products (n = 8) derived from all CBU groups (A–D) were used. APB PRP samples (n = 8) were obtained from Evagelismos Hospital. Expired APB units were used for the production of PRP. APB collection was performed from healthy donors, following the Greek regulatory procedures for blood donation.

Initially, total protein content of PRP derived either from CBUs (n = 4) or APB (n = 4) was quantified with Bradford assays. Then, the samples were diluted with urea buffer (8 M urea, 50 mM NH$_4$HCO$_3$, Sigma-Aldrich, Darmstadt, Germany) to reach final volume of 20 μL. Reduction using 10 mM dithioerythritol (DTE, Sigma-Aldrich, Darmstadt, Germany) and alkylation with 40 mM iodoacetimide (Sigma-Aldrich, Darmstadt, Germany) was performed, followed by dilution with 50 mM NH$_4$HCO$_3$ (Sigma-Aldrich, Darmstadt, Germany) until reaching a final volume of 90 μL. Overnight trypsinization was performed in each sample with enzyme-protein ratio of 1:100. The next day, 0.1% formic acid (Sigma-Aldrich, Darmstadt, Germany) was added. Finally, the samples were desalted by zip-tip and vacuum dried (SpeedVac vacuum concentrators, Thermo Fisher Scientific, Waltham, MA, USA).

The samples were reconstituted to a final protein concentration of 0.5 μg/mL and analyzed by LC/MRM. A total number of 31 proteins were tested for with this approach (Table S2). The protein determination was performed according to previous data in the literature with some modifications [11].

Skyline software and Peptideatlas repository were used for the identification of the proteotypic peptides of PRP samples. All chromatograms obtained from LC-MRM were visually inspected. The sum of peak areas of two to four transitions per peptide was used to calculate the signal intensity for the selected growth factors. A detailed list of MRM transitions is provided as supplementary material (Table S4).

Further analysis involved the classification of the identified proteins using the Panther classification system (www.pantherdb.org).

2.4. Liquid Chromatography/Multiple Reaction Monitoring Setup

Agilent 1200 series (Agilent Technologies, Inc., Santa Clara, CA, USA) coupled with CS18 nano-column (150 mm × 75 µm, particle size 5 µm) was used for LC. Peptide separation and elution was performed using a 40 min 5–45% ACN/water 0.1% FA gradient at a flow rate of 300 nL/min. Six microliters of each sample (corresponding to 3 µg of total protein content) were injected. AB/MDS Sciex 4000 QTRAP coupled with a nanoelectrospray ionization source was used for the tryptic peptide analysis. The mass spectrometer was operated in MRM mode, with the first (Q1) and third quadrupole (Q3) at 0.7 unit mass resolution. Two to four transitions of each peptide were recorded. Optimum collision energies for each transition were automatically calculated by the Skyfline software.

2.5. Contamination Validation Study

All CBUs, APB samples and produced PRP samples were tested for HIV, HBV, HGV, HTLV-I/II, CMV, HCV, HAV, WNV, T Pallidum, syphilis, aerobic anaerobic bacteria. Furthermore, PRP samples derived from both sources were evaluated for Mycoplasma contamination and endotoxin level.

2.6. Evaluation of Fibrin Gel

The produced fibrin gels ($n = 10$) from each group were further evaluated. For this purpose, the time needed for the development of fibrin gel, after the addition of 10% calcium gluconate, was determined. In addition, the mean fibrin gel area was determined with the use of a digital caliper (Mitutoyo, Radionics, Ltd., Dublin, Ireland). Finally, the degradation time of fibrin gel was also estimated. The estimation of fibrin gel stability time was performed at room temperature (RT, 21–25 °C) and 37 °C. The fibrin gel stability time corresponded to hours (h). At RT, the fibrin gels from all groups were placed in open Petri dishes (Sigma-Aldrich, Darmstadt, Germany), until complete gel degradation. For the second measurements, the fibrin gels were placed in 50 mL falcon tubes (BD falcon tubes, Corning, NY, USA), and immersed into a water bath (WNB 10, Memmert, Gmbh, Schwabach, Germany) at 37 °C, until complete gel degradation.

2.7. Statistical Analysis

Statistical analysis was performed using GraphPad Prism v6 (GraphPaD Software, San Diego, CA, USA). Comparisons of PRP data were performed with the unpaired Kruskal Wallis test. Statistically significant differences were obtained when the p value was less than 0.05. Data were presented as medians and mean ± standard deviation (SD). Pie charts of protein classifications were produced with Microsoft Excel 2016 (Microsoft Office for Windows 2016, Redmond, Washington, DC, USA).

3. Results

3.1. Platelet Rich Plasma Data Analysis

A total number of 200 CBUs that did not fulfill the minimum criteria for processing and banking of the HCBB (Table S1) were used for the evaluation of fibrin gel production. The collection of CBUs was started in October of 2018 and ended in April of 2019. The CBUs were divided into 4 groups according to their volume (including the anticoagulant). The characteristics of each group prior and

PRP production are represented in Table 1. The initial cord blood volumes in groups A to D, were 73 ± 6 mL, 97 ± 9 mL, 127 ± 11 mL and 130 ± 12 mL, respectively (Table 1).

The mean PLT concentration and the initial PLT number of all groups are represented in Table 1. Moreover, the initial PLT number was correlated positively with the initial CBU volume, as shown in Figure 1A.

After the centrifugation steps, the final volume of PRP in groups A to D was 6 ± 1 mL, 7 ± 2 mL, 8 ± 2 mL and 8 ± 2 mL, respectively (Figure 1). The median PLT number in PRP of groups A to D, was 17.3×10^9, 21.5×10^9, 22.6×10^9 and 22.7×10^9 respectively (Table 1, Figure 1). The mean of WBCs and RBCs in PRP of all groups was less than 4×10^3/µL and 1×10^6/µL, respectively (Figure 1, Table S3). Accordingly, the median recovery of PLTs in PRP groups A to D, was 30%, 34%, 48% and 48%, respectively (Table 2). Statistically significant differences resulted regarding the concentration ($p < 0.001$) and number ($p < 0.001$) of PLTs, as well as volume ($p < 0.001$) and recovery ($p < 0.001$) after the PRP production, between all groups.

Table 1. Characteristics of cord blood units prior processing, distributed in four groups. Statistically significant differences were observed in initial platelet (PLT) number between all groups ($p < 0.05$). [a] The values represented are medians. [b] The values are represented as mean ± sd.

n = 40/Group	Group A (<81 mL)	Group B (82–110 mL)	Group C (111–148 mL)	Group D-Pooled (111–148 mL)	p-Value
Initial Net Weight (g)	(78) [a] 77 ± 7 [b]	(103) 103 ± 3	(131) 134 ± 12	(138) 138 ± 11	-
Initial Net Volume (mL)	(74) 73 ± 6	(97) 97 ± 9	(124) 127 ± 11	(128) 130 ± 12	-
PLTs ($\times 10^3$/µL)	(213) 214 ± 43	(223) 223 ± 35	(187) 182 ± 55	(171) 184 ± 39	ns
Initial PLTs count ($\times 10^6$)	(17298) 16831 ± 3601	(21510) 22687 ± 3791	(22580) 23100 ± 3659	(22672) 23850 ± 3657	<0.05

Table 2. Characteristics of platelet rich plasma (PRP) distributed in the four groups. Statistically significant differences were observed in final volume ($p < 0.001$), PLT concentration ($p < 0.001$), PLT number ($p < 0.001$) and recovery ($p < 0.001$) between all groups. [a] The values represented are medians. [b] The values are represented as mean ± sd.

n = 40/Group	Group A (<81 mL)	Group B (82–110 mL)	Group C (111–148 mL)	Group D-Pooled (111–148 mL)	p-Value
PRP Volume (mL)	(6) [a] 6 ± 1 [b]	(7) 7 ± 2	(8) 8 ± 2	(8) 8 ± 2	<0.001
Final PLTs ($\times 10^3$/µL)	(702) 761 ± 246	(1142) 1107 ± 367	(1366) 1382 ± 425	(1352) 1388 ± 345	<0.001
Final PLTs count ($\times 10^6$)	(4407) 5331 ± 2355	(7151) 8120 ± 3528	(10568) 11066 ± 3260	(10523) 11345 ± 3496	<0.001
Recovery (%)	(30) 32 ± 11	(34) 35 ± 10	(48) 48 ± 10	(48) 48 ± 11	<0.001

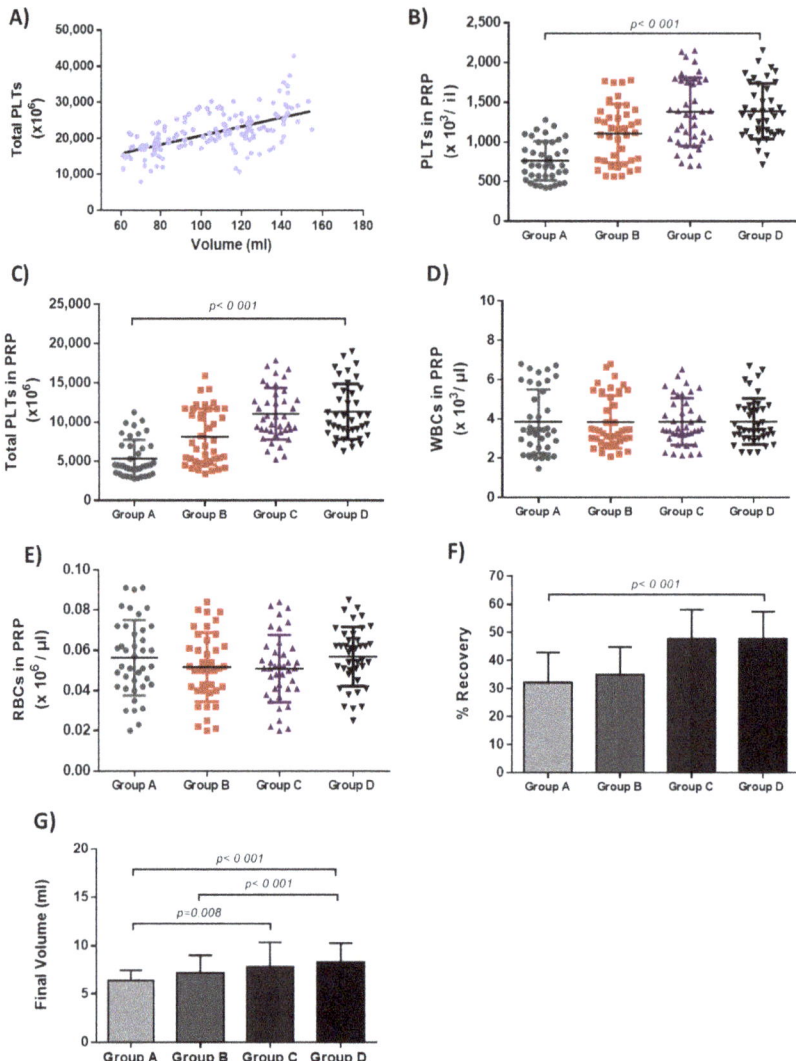

Figure 1. Characteristics of PRP samples from all groups. (**A**) Correlation between initial PLT number and initial cord blood (CB) volume, $R^2 = 0.3362$. (**B**) PLT concentration in PRP samples derived from all groups. Statistically significant differences were observed in PLT concentration between groups A–D ($p < 0.001$). (**C**) Total number of PLTs in PRP samples derived from all groups. Statistically significant differences were observed in total PLT number between groups A–D ($p < 0.001$). (**D**) White blood cell (WBC) concentration in PRP samples derived from all groups. (**E**) Red blood cell (RBC) concentration in PRP samples derived from all groups. (**F**) PLT recovery in PRP samples. Statistically significant differences were observed in PLT recovery between all groups ($p < 0.001$). (**G**) Final volume of PRP samples. Statistically significant differences were observed between group A and group C ($p = 0.008$) and D ($p < 0.001$), and between group B and group D ($p < 0.001$).

3.2. Protein Identification

LC/MRM technology was used for the identification of proteins in PRP from CBUs and APB. A number of 25 proteins, which corresponded to 81% of the proposed proteins, were successfully identified both in PRP from CBUs, and APB samples (Table 3 and Table S4). The correspondingly identified peptide ions of each proteins are represented in Table S4. Identical proteins were identified in PRP and APB samples.

Furthermore, the identified proteins were classified using the Panther classification system. Based on their biological function, 11% of proteins were transferases, 21%, immunity/defense proteins, 26% receptors and 42% signaling molecules (Figure 2 and Table S5).

Table 3. List of protein identifications in adult peripheral blood (APB) and cord blood (CB) PRP.

No.	Growth Factors	Entry Name	Accession Number
1	Tumor Neccrosis Factor A (TNF A)	TNFA_HUMAN	P01375
2	Interleukin-1A (IL-1A)	IL1A_HUMAN	P01583
3	Interleukin-1B (IL-1B)	IL1B_HUMAN	P01584
4	Interleukin-2 (IL-2)	IL2_HUMAN	P60568
5	Interleukin-6 (IL-6)	IL6_HUMAN	P05231
7	Interleukin-8 (IL-8)	IL8_HUMAN	P10145
9	Tumour necrosis factor receptor type 1-associated DEATH domain protein (TRADD)	TRADD_HUMAN	Q15628
10	Interleukin-1 Receptor (IL-1R)	IL1R1_HUMAN	P14778
11	Interleukin-2 Receptor (IL-2GR)	IL2RG_HUMAN	P31785
12	Interleukin-6 Receptor (IL-6R)	IL6RA_HUMAN	P08887
15	Interleukin-10 Receptor 1 (IL-10R1)	I10R1_HUMAN	Q13651
16	Interleukin-10 Receptor 2 (IL-10R2)	I10R2_HUMAN	Q08334
17	Vascular Endothelial Growth Factor A (VEGF-A)	VEGFA_HUMAN	P15692
18	Vascular Cell Adhesion protein 1 precursor (VCAM-1)	VCAM1_HUMAN	P19320
19	Intracellular Cell Adhesion protein 1 precursor (ICAM-1)	ICAM1_HUMAN	P05362
20	Platelet Derived Growth Factor AA (PDGF-AA)	PDGFA_HUMAN	P04085
21	Transforming Growth Factor B1 (TGF-B1)	TGFB1_HUMAN	P01137
22	Fibroblast Growth Factor 2 (FGF2)	FGF2_HUMAN	P09038
23	C-C motif chemokine receptor 1 (CCR1)	CCR1_HUMAN	P32246
24	Transforming Growth Factor-B receptor 1 (TGF-B R1)	TGFR1_HUMAN	P36897
25	Transforming Growth Factor-B receptor 2 (TGF-B R2)	TGFR2_HUMAN	P37173

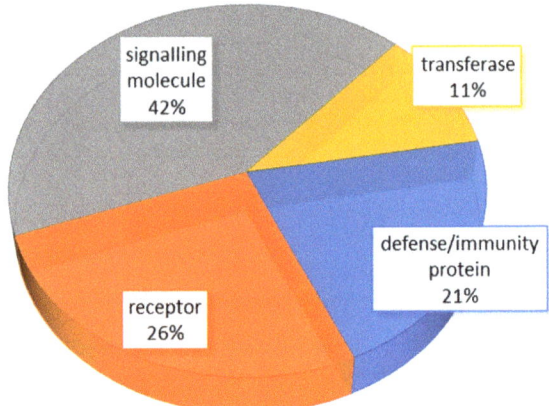

Figure 2. Classification of identified proteins based on the Panther classification system.

3.3. Platelet Rich Plasma and Adult Peripheral Blood Validation Tests

All CBUs, APB and PRP samples were negative for bacterial or viral contamination. All samples were negative for aerobic or anaerobic bacteria according to the BacT/Alert system, which were further confirmed by blood and Sabouraud agar (Table S6). In addition, PRP and APB samples were negative for HIV I/II, HBV, HGV, HTLV-I/II, CMV, HCV, HAV, WNV, and for *Treponema pallidum* and *Trypanosoma cruzi* (Table S7). The endotoxin level in PRP samples derived from both sources, was less than 2.5 EU/mL, while no mycoplasma contamination was detected (Table S7).

3.4. Evaluation of Fibrin Gel Production

The PRP from all groups (A to D) were solidified successfully after the addition of calcium gluconate (Figure 3). The developed fibrin was characterized by a gelatinous form, which was lifted without breaking from the petri dish (Figure 3).

The mean time needed for the fibrin development of groups A to D, was 21 ± 1 min, 22 ± 2 min, 22 ± 2 min and 22 ± 1 min, respectively (Figure 3). Furthermore, the mean covered surface of each fibrin gel was estimated. Specifically, the fibrin gel surface of groups A to D, was 5 ± 1 cm^2, 6 ± 1 cm^2, 12 ± 1 cm^2 and 12 ± 1 cm^2, respectively (Figure 3). Statistically significant differences were observed only in fibrin gel produced surface between groups A and B, with groups C ($p < 0.001$) and D ($p < 0.001$). Further evaluation involved the determination of fibrin gel stability time at RT and 37 °C. The mean degradation time of fibrin gels at RT in groups A–D was 1.3 ± 0.2 h, 2 ± 0.3 h, 4 ± 0.4 h and 4 ± 0.3 h, respectively (Table 4). The mean degradation time of fibrin gels at 37 °C in groups A–D, was 3 ± 0.3 h, 4 ± 0.4 h, 7 ± 0.4 h and 7 ± 0.4 h, respectively (Table 4). Statistically significant differences in fibrin gel degradation time between groups A–D, at RT ($p < 0.001$) and 37 °C ($p < 0.001$), were found.

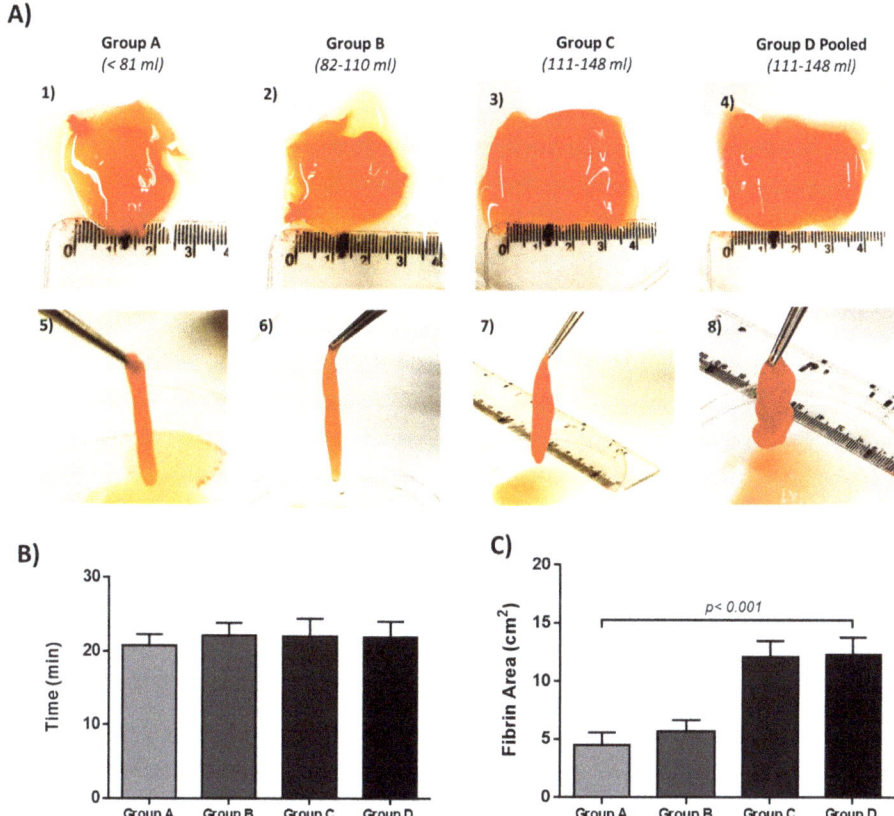

Figure 3. Characteristics of fibrin gel development. (**A**) Fibrin gel obtained from group A (A1, A5), group B (A2, A6), group C (A3, A7) and group D (A4, A8). (**B**) Time needed for the development of fibrin gel derived from all groups. (**C**) Fibrin gel area obtained from all groups. Statistically significant differences were observed in fibrin gel area between all groups ($p < 0.001$).

Table 4. Fibrin gel stability time at 25 °C and 37 °C. Statistically significant differences in fibrin gel stability time at 25 °C ($p < 0.001$) and 37 °C ($p < 0.001$) were observed between all groups.

Fibrin Gel Degradation Temperature	Group A (h)	Group B (h)	Group C (h)	Group D (h)	p-Value
RT (21–25 °C)	1.3 ± 0.2	2 ± 0.3	4 ± 0.4	4 ± 0.3	<0.001
37 °C	3 ± 0.3	4 ± 0.4	7 ± 0.4	7 ± 0.4	<0.001

4. Discussion

Fibrin gel has been used widely as a tool for tissue engineering and regenerative medicine approaches. At this point, fibrin gel and PRP products derived either from autologous or allogeneic sources have been used for the treatment of skin ulcers, bone and cartilage repair, and even more prominently in cosmetics surgeries [7]. Although these autologous medicinal products are considered microbiologically and virally safe, a number of practical limitations may hamper their clinical use [7,17]. In cases where repeated blood collections must be performed from patients, including the elderly, neonates and patients with cardiovascular diseases or hematological malignancies, the autologous products may be clinically inappropriate.

Due to these drawbacks, the allogeneic fibrin gel may be used in these categories of patients. Allogeneic fibrin gel mostly is produced from expired blood units or PLT units after platelet apheresis [20,21]. However, expired blood units may increase the risk of contamination, thus the products may be characterized by low clinical effect. On the other hand, discarded CBUs from public CBBs could represent a significant source for production of fibrin gel [6,18]. It is estimated that over of 80% of the delivered CBUs in CBBs are discarded due to their low stem cell content and collected volume, thus are not appropriate for a hematopoietic stem cell transplantation.

The aim of this study was the standardization of fibrin gel production from discarded CBUs, in order to develop a highly reproducible method, irrespective of their initial volumes. Thus, low volume CBUs could be combined and used for fibrin gel production. For this purpose, a number of 200 CBUs that were delivered to HCBB, and did not meet the minimum criteria for processing and banking, were used.

The results of this study showed that final PLT recovery is dependent on CBU volume and initial PLT number. High volume CBUs of group C and D were characterized by better initial PLT number, recovery rates and final PLT yields, when compared to low volume CBUs. Although low volume CBUs represent the majority of the received CBUs at the HCBB, these units were difficult to handle in the laboratory setting. Due to the low volume, there was a risk in obtaining high yields of WBCs and RBCs in the final PRP product. For this reason, the extraction process in low volume CBUs was not performed appropriately, thus resulting to lower recovery rates and final PLT concentration compared to the high volume CBUs. In our study, the PLT concentration of low volume CBUs ($<800 \times 10^6$ PLTs/μL) in the final product, was below the proposed criteria for cord blood PRP production [6], affecting in this way, its potential therapeutic use. In order to avoid the destruction of a high number of low volume units, pooling of CBUs with the same blood group was performed.

Pooled CBUs (group D) were characterized by similar results regarding the recovery rates, initial and final PLT concentration, and total number as group C. Furthermore, all groups were characterized by no statistically significant differences in concentration of WBCs and RBCs in the final product. Then, we tried to compare our results with a multicenter clinical grade study that was accomplished by Rebulla et al [6]. In this study, CBUs with volume of 97.6 ± 20 mL were used, and resulted in PLT recovery over of 47%. However, in our study, this recovery rate was achieved only from high volume CBUs (>111 mL). This minor discrepancy in the results between these two studies, might be explained by the use of slightly different procedures. In addition, that difference may disappear if we include more CBUs in our validation study. In the study of Rebulla et al. [6], a number of 1080 CBUs were used, whereas in our study, only 200 CBUs in total were included. Moreover, both studies showed that proper PRP production could not be performed from low volume CBUs (<97 mL). All PRP samples that were produced with the proposed protocol were negative for HIV I/II, HBV, HGV, HTLV-I/II, CMV, HCV, HAV, WNV, and for *T. pallidum* and *T. cruzi*. Specifically, the endotoxin level in PRP samples was less than 2.5 EU/mL.

The next step of this study, involved the proteomic analysis of PRP products. In order to verify that PRP derived from CBUs was not different in protein content from PRP of other sources, LC/MRM technology was applied. Proteomic analysis was performed in CB PRP products and compared with the proteomic profile of PRP derived from APB. In this way, 25 proteins of the 31 initially assessed, were successfully identified both in APB and CB PRP products, indicating that both products were similar in their protein content. The identified proteins were classified as transferases (11%), defense/immunity proteins (21%), receptors (26%) and signaling molecules (42%). No discrepancy in protein identification was observed between APB and CB fibrin gel. This fact is important, regarding the therapeutic potential of fibrin gel. In the study of Stokhuijen et al. [22], differences between PLTs from APB and CBUs were observed only for intracellular proteins which belonged to metabolic pathways, and were not adhesive integrins or glycoproteins. This in turn indicated that either CB or APB PLT derived products can act in the same way, without any significant alteration to their therapeutic result. Indeed, Tadini et al. [23], Janmey et al. [7] and Perseghin et al. [24] have confirmed the beneficial effect of CB derived fibrin gel and PRP in the treatment of skin lesions produced by dystrophic epidermolysis bullosa and chronic wounds. Furthermore, the therapeutic properties of fibrin gel may be useful in bone and muscle tissue regeneration. Fibrin gel can be combined efficiently with acellular matrices. Indeed, Aulino et al. described a novel approach for muscle and bone regeneration through the use of decellularized tibialis anterior [25]. In this context, and knowing that specific growth factors, such as TGF-β1, FGF, VEGF, PDGF, and cytokines, can guide better the migration and differentiation of stem cells, the use of fibrin gel might be a beneficial approach [25]. Fibrin gel contains significant amounts of growth factors and cytokines, with defined viscoelastic properties, thus it can be used separately or in combination with various matrices in order to be used as a tool for regenerative medicine applications.

Based on their biological function, these proteins play significant role in tissue regeneration. Among them, growth factors including TGF-β1, FGF, VEGF-A, PDGF-A have been reported in literature for their potential contribution in signaling pathways responsible for cell activation, proliferation and differentiation. Moreover, that growth factor repertoire has been implicated in regulation of phosphatase and tensin homolog (PTEN), a significant signaling protein which both reduces the hypertrophic scar tissue and promotes fibroblast transdifferentiation [25]. Furthermore, the growth factors can promote, via MAP/ERK and Wnt singalling pathways, the regeneration of osteocytes and chondrocytes at the site of the injury [26]. In addition, the identified cytokine receptors and adhesion molecules can potentially contribute even more in cell proliferation and regeneration of damaged tissues [11].

In the current study, no matrix metaloproitenases (MMPs) were able to be identified in all PRP samples using LC/MRM. MMPs are enzymes responsible for the destruction of extracellular matrix components during the fibrosis process. These enzymes are also used by cancer cells in order to perform tumor metastasis [27]. The presence of MMPs in the PRP product could act negatively in the tissue regeneration process. The fact that no MMPs were identified in PRP from both sources, supports further its beneficial effect in tissue regeneration and wound healing. Up to now, the presence of MMPs in the PRP is still under debate. Several studies are in accordance with our findings [4,28], but evidence from other research groups confirms the presence of MMPs in PRP samples [29,30]. This phenomenon might be explained due to the different preparation protocols of PRP that are used by the research groups, worldwide. More work needs to be performed in this field in order to obtain safer results.

Finally, PRP samples from all groups of this study were solidified successfully, thus producing the fibrin gel. In less than 30 min at 37 °C, the fibrin gel was produced with the addition of appropriate volume of calcium gluconate. In addition, high volume CBUs of group C and D were able to produce wider surface gel area compared to the low volume units of group A and B. In this way, a single fibrin a gel could be administrated in a wound area of 5–12 cm^2, while a greater number of fibrin gels could be used to wider lesions. No significant alterations in fibrin gel production were reported in the literature when thrombin or batroxobin instead of calcium glugonate, have been used [4]. In addition, fibrin gels derived from group C and D, needed greater time to be degraded when compared to those of group A and B. Owing to that, the fibrin gel content, including the growth factors and the cytokines, could be present for longer time at the site of injury, providing greater regenerative potential.

The above data indicated that CBUs with initial volumes greater than 82 mL could be used, potentially, for fibrin gel production. However, the efficacy of fibrin gel production may be significantly improved when pooled CBUs are used. Before pooling, the initial PLT concentration in CBUs must be determined. Only CBUs with concentration greater than 150×10^3 PLTs/μL, can be pooled. Therefore, it is estimated that the concentration in fibrin gel will be greater than 800×10^3 PLTs/μL, which has been described as a significant therapeutic dose for regenerative medicine approaches. Ideally, the PRP after its production from CBUs can be stored at −80 °C. On demand, the PRP could be thawed in a water bath at 37 °C, followed by the addition of calcium gluconate, in order to form the fibrin gel. Under that model, a PRP storage bank could be developed, where the produced fibrin gel, with the same blood group with the patients, could be administrated for regenerative medicine applications, such as wound healing, extended skin lesions, and bone and muscle regeneration. Regarding the determination of skin lesions or the establishment of bone and muscle damage, bioimpedance detection could be applied by the physicians. Bioimpedance analysis is a powerful diagnostic tool than can be used for the classification of skin lesions and mucosa damage [31]. Recently, in the study of Tatullo et al. [31], the bioimpedance detection system was used in oral lichen planus lesions that can potentially lead to malignant transformations. In that context, and by determining accurately the tissue damage in patients, fibrin gel could be administrated in different ways in order to achieve the best outcome.

Fibrin gel could be a useful tool for tissue regeneration, which can be produced with a different amount of stiffness, viscoelastic properties and degradation rates [1]. They can be achieved by modifying the key elements of solidification process, such as the concentration of calcium gluconate, batroxobin, sodium chloride, or the divalent cations of Ca and Mg [1]. In this context, the gel which will be used in various skin lesions, may be characterized by high bioabsorption rate. On the other hand, in bone tissue engineering, fibrin gels with increased stiffness would be applied. Furthermore, when fibrin gel could be used as a cell carrier, it must be characterized by a specific pore size, providing a suitable cellular microenvironment. In order to achieve that, fibrin gel could be combined with various chemical compounds, such as nanosilicates. Kerativitayanan et al. [32] presented a possible production of osteoinductive and elastomeric scaffolds with specific pore sizes by combining polyglycerol sebacate and nanoslicates.

Future experiments will include a greater number of CBUs, in order to obtain more valid conclusions regarding the process reproducibility

5. Conclusions

Fibrin gel is naturally derived hydrogel that can be developed in a few minutes and is characterized by unique viscoelastic properties. Its production cost is relatively low, as has been indicated in literature [6]. The potential use of low volume, discarded CBUs for fibrin gel production is of major importance. Worldwide, public CBBs accredited by the Foundation for the Accreditation of Cellular Therapy (FACT) are economically burdened by the great number of low volume CBUs that must be discarded. The results of our study indicate that the low volume CBUs can be used for fibrin gel development, which could be used as a tool by physicians in regenerative medicine approaches. With that in mind, a PRP storage bank can be established, where the PRP samples can be stored over a long time period, and thawed on demand, in order to form the fibrin gel. The produced fibrin gel could be a useful tool for physicians in various regenerative medicine approaches.

Supplementary Materials: The following are available online at http://www.mdpi.com/2306-5354/6/3/66/s1: Table S1: Acceptable range of results outlined by the Hellenic Cord Blood Bank for processing and storage of cord blood units. Table S2: List of proteins for identification with LC/MRM. Table S3: WBCs and RBCs prior and after processing. * The values are represented as median. † The values are represented as mean ± sd. Table S4: List of parent ions and fragments used in MRM analysis of PB-PL and CB-PL growth factors. Table S5: Protein classification based on their biological functions. Table S6: Validation test for CBUs and APB samples. Table S7: Validation test for PRP derived from CBUs and APB samples.

Author Contributions: P.M. (first author) and I.G. (equal first author) carried out the whole experimental procedure of this study and prepared the manuscript. In addition, P.M. performed the statistical analysis. Z.D., E.P., V.G. and E.G. contributed in the experimental procedure. M.M. and J.Z. performed the proteomic analysis. A.V. supervised the whole proteomic analysis. C.S.-G. supervised and approved the overall study. E.M. supervised and approved the study.

Funding: This research received no funding.

Conflicts of Interest: The authors declare no conflict of interest.

References

1. Murphy, K.C.; Leach, J.K. A reproducible, high throughput method for fabricating fibrin gels. *BMC Res. Notes* **2012**, *5*, 423. [CrossRef] [PubMed]
2. Shaikh, F.M.; Callanan, A.; Kavanagh, E.G.; Burke, P.E.; Grace, P.A.; McGloughlin, T.M. Fibrin: A natural biodegradable scaffold in vascular tissue engineering. *Cells Tissues Organs* **2008**, *188*, 333–346. [CrossRef] [PubMed]
3. Tsachiridi, M.; Galyfos, G.; Andreou, A.; Sianou, A.; Sigala, F.; Zografos, G.; Filis, K. Autologous Platelet-Rich Plasma for Nonhealing Ulcers: A Comparative Study. *Vasc. Spec. Int.* **2019**, *35*, 22–27. [CrossRef] [PubMed]
4. Parazzi, V.; Lavazza, C.; Boldrin, V.; Montelatici, E.; Pallotti, F.; Marconi, M.; Lazzari, L. Extensive Characterization of Platelet Gel Releasate from Cord Blood in Regenerative Medicine. *Cell Transplant.* **2015**, *24*, 2573–2584. [CrossRef] [PubMed]
5. Astori, G.; Amati, E.; Bambi, F.; Bernardi, M.; Chieregato, K.; Schäfer, R.; Sella, S.; Rodeghiero, F. Platelet lysate as a substitute for animal serum for the ex-vivo expansion of mesenchymal stem/stromal cells: Present and future. *Stem Cell Res. Ther.* **2016**, *7*, 93. [CrossRef] [PubMed]
6. Rebulla, P.; Pupella, S.; Santodirocco, M.; Greppi, N.; Villanova, I.; Buzzi, M.; De Fazio, N.; Grazzini, G. Multicentre standardisation of a clinical grade procedure for the preparation of allogeneic platelet concentrates from umbilical cord blood. *Blood Transfus.* **2016**, *14*, 73–79. [PubMed]
7. Janmey, P.; Winer, J.P.; Weisel, J.W. Fibrin gels and their clinical and bioengineering applications. *J. R. Soc. Interface* **2009**, *6*, 1–10. [CrossRef]
8. Blombäck, B.; Bark, N. Fibrinopeptides and fibrin gel structure. *Biophys. Chem.* **2004**, *112*, 147–151. [CrossRef]
9. Qiao, J.; An, N.; Ouyang, X. Quantification of growth factors in different platelet concentrates. *Platelets* **2017**, *28*, 774–778. [CrossRef]
10. Kobayashi, E.; Flückiger, L.; Fujioka-Kobayashi, M.; Sawada, K.; Sculean, A.; Schaller, B.; Miron, R.J. Comparative release of growth factors from PRP, PRF, and advanced-PRF. *Clin. Oral Investig.* **2016**, *20*, 2353–2360. [CrossRef]
11. Christou, I.; Mallis, P.; Michalopoulos, E.; Chatzistamatiou, T.; Mermelekas, G.; Zoidakis, J.; Vlahou, A.; Stavropoulos-Giokas, C. Evaluation of Peripheral Blood and Cord Blood Platelet Lysates in Isolation and Expansion of Multipotent Mesenchymal Stromal Cells. *Bioengineering* **2018**, *5*, 19. [CrossRef] [PubMed]
12. Heijnen, H.F.; Schiel, A.E.; Fijnheer, R.; Geuze, H.J.; Sixma, J.J. Activated platelets release two types of membrane vesicles: Microvesicles by surface shedding and exosomes derived from exocytosis of multivesicular bodies and alpha-granules. *Blood* **1999**, *94*, 3791–3799. [PubMed]
13. Etulain, J. Platelets in wound healing and regenerative medicine. *Platelets* **2018**, *29*, 556–568. [CrossRef] [PubMed]
14. Weisel, J.W. Fibrinogen and fibrin. *Adv. Protein Chem.* **2005**, *70*, 247–299. [PubMed]
15. Weisel, J.W. Structure of fibrin: Impact on clot stability. *J. Thromb. Haemost.* **2007**, *5*, 116–124. [CrossRef] [PubMed]
16. Weisel, J.W. Which knobs fit into which holes in fibrin polymerization. *J. Thromb. Haemost.* **2007**, *5*, 2340–2343. [CrossRef] [PubMed]
17. Parazzi, V.; Lazzari, L.; Rebulla, P. Platelet gel from cord blood: A novel tool for tissue engineering. *Platelets* **2010**, *21*, 549–554. [CrossRef] [PubMed]
18. Panagouli, E.; Dinou, A.; Mallis, P.; Michalopoulos, E.; Papassavas, A.; Spyropoulou-Vlachou, M.; Meletis, J.; Angelopoulou, M.; Konstantopoulos, K.; Vassilakopoulos, T.; et al. Non-Inherited Maternal Antigens Identify Acceptable HLA Mismatches: A New Policy for the Hellenic Cord Blood Bank. *Bioengineering* **2018**, *5*, 77. [CrossRef] [PubMed]
19. Maynard, D.M.; Heijmen, H.F.G.; Horne, M.K.; White, J.G.; Gahl, W.A. Proteomic analysis of platelet α-granules using mass spectrometry. *J. Thromb. Haemost.* **2007**, *5*, 1945–1955. [CrossRef] [PubMed]
20. Piccin, A.; Di Pierro, A.M.; Canzian, L.; Primerano, M.; Corvetta, D.; Negri, G.; Mazzoleni, G.; Gastl, G.; Steurer, M.; Gentilini, I.; et al. Platelet gel: A new therapeutic tool with great potential. *Blood Transfus.* **2017**, *15*, 333–340. [PubMed]
21. Anitua, E.; Prado, R.; Orive, G. Allogeneic Platelet-Rich Plasma: At the Dawn of an Off-the-Shelf Therapy? *Trends Biotechnol.* **2017**, *35*, 91–93. [CrossRef] [PubMed]

22. Stokhuijzen, E.; Koornneef, J.M.; Nota, B.; van den Eshof, B.L.; van Alphen, F.P.J.; van den Biggelaar, M.; van der Zwaan, C.; Kuijk, C.; Mertens, K.; Fijnvandraat, K.; et al. Differences between Platelets Derived from Neonatal Cord Blood and Adult Peripheral Blood Assessed by Mass Spectrometry. *J. Proteome Res.* **2017**, *16*, 3567–3575. [CrossRef] [PubMed]
23. Tadini, G.; Guez, S.; Pezzani, L.; Marconi, M.; Greppi, N.; Manzoni, F.; Rebulla, P.; Esposito, S. Preliminary evaluation of cord blood platelet gel for the treatment of skin lesions in children with dystrophic epidermolysis bullosa. *Blood Transfus.* **2015**, *13*, 153–158. [PubMed]
24. Perseghin, P.; Sciorelli, G.; Belotti, D.; Speranza, T.; Pogliani, E.M.; Ferro, O.; Gianoli, M.; Porta, F.; Paolini, G. Frozen-and-thawed allogeneic platelet gels for treating postoperative chronic wounds. *Transfusion* **2005**, *45*, 1544–1546. [CrossRef] [PubMed]
25. Aulino, P.; Costa, A.; Chiaravalloti, E.; Perniconi, B.; Adamo, S.; Coletti, D.; Marrelli, M.; Tatullo, M.; Teodori, L. Muscle Extracellular Matrix Scaffold Is a Multipotent Environment. *Int. J. Med. Sci.* **2015**, *12*, 336–340. [CrossRef] [PubMed]
26. Liu, Y.; Li, Y.; Li, N.; Teng, W.; Wang, M.; Zhang, Y.; Xiao, Z. TGF-β1 promotes scar fibroblasts proliferation and transdifferentiation via up-regulating MicroRNA-21. *Sci. Rep.* **2016**, *6*, 32231. [CrossRef]
27. Deryugina, E.I.; Quigley, J.P. Matrix metalloproteinases and tumor metastasis. *Cancer Metastasis Rev.* **2006**, *25*, 9–34. [CrossRef]
28. Longo, V.; Rebulla, P.; Pupella, S.; Zolla, L.; Rinalducci, S. Proteomic characterization of platelet gel releasate from adult peripheral and cord blood. *Proteom. Clin. Appl.* **2016**, *10*, 870–882. [CrossRef]
29. Pifer, M.; Maerz, T.; Baker, K.C.; Anderson, K. Matrix metalloproteinase content and activity in low-platelet, low-leukocyte and high-platelet, high-leukocyte platelet rich plasma (PRP) and the biologic response to PRP by human ligament fibroblasts. *Am. J. Sports Med.* **2014**, *42*, 1211–1218. [CrossRef]
30. Kardos, D.; Simon, M.; Vácz, G.; Hinsenkamp, A.; Holczer, T.; Cseh, D.; Sárközi, A.; Szenthe, K.; Bánáti, F.; Szathmary, S.; et al. The Composition of Hyperacute Serum and Platelet-Rich Plasma Is Markedly Different despite the Similar Production Method. *Int. J. Mol. Sci.* **2019**, *20*, 721. [CrossRef]
31. Tatullo, M.; Marrelli, M.; Amantea, M.; Paduano, F.; Santacroce, L.; Gentile, S.; Scacco, S. Bioimpedance Detection of Oral Lichen Planus Used as Preneoplastic Model. *J. Cancer* **2015**, *6*, 976–983. [CrossRef]
32. Kerativitayanan, P.; Tatullo, M.; Khariton, M.; Joshi, P.; Perniconi, B.; Gaharwar, A.K. Nanoengineered Osteoinductive and Elastomeric Scaffolds for Bone Tissue Engineering. *ACS Biomater. Sci. Eng.* **2017**, *3*, 590–600. [CrossRef]

© 2019 by the authors. Licensee MDPI, Basel, Switzerland. This article is an open access article distributed under the terms and conditions of the Creative Commons Attribution (CC BY) license (http://creativecommons.org/licenses/by/4.0/).

Article

Administration of Adipose Derived Mesenchymal Stem Cells and Platelet Lysate in Erectile Dysfunction: A Single Center Pilot Study

Vassilis Protogerou [1,2], Efstathios Michalopoulos [3,*], Panagiotis Mallis [3], Ioanna Gontika [3], Zetta Dimou [3], Christos Liakouras [2], Catherine Stavropoulos-Giokas [3], Nikolaos Kostakopoulos [2], Michael Chrisofos [2] and Charalampos Deliveliotis [4]

1. Department of Anatomy and Surgical Anatomy, Medical School of Athens, National and Kapodistrian University, 12462 Athens, Greece
2. 2nd Urological Department, Attikon Hospital, Medical School of Athens, National and Kapodistrian University, 12462 Athens, Greece
3. Hellenic Cord Blood Bank, Biomedical Research Foundation Academy of Athens, 4 Soranou Ephessiou Street, 11527 Athens, Greece
4. 2nd Urological Department, Medical School of Athens, National and Kapodistrian University, 12462 Athens, Greece
* Correspondence: smichal@bioacademy.gr; Tel.: +30-210-597331; Fax: +30-210-6597345

Received: 16 January 2019; Accepted: 28 February 2019; Published: 5 March 2019

Abstract: Erectile dysfunction (ED) affects more than 30 million men; endothelial dysfunction plays a significant role in EDs pathogenesis. The aim of this study was to administer mesenchymal stem cells (MSC) derived from adipose tissue and platelet lysate (PL) into patients with erectile dysfunction. This pilot study enrolled eight patients with diagnosed ED. Patients enrolled were suffering from organic ED due to diabetes melitus, hypertension, hypercholesterolaemia, and Peyronie disease. The patients were distributed in 2 groups. Patients in group A received adipose derived mesenchymal stem cells (ADMSC) resuspended in PL while patients in group B received only PL. ADMSCs were isolated from patients' adipose tissue and expanded. In addition, blood sampling was obtained from the patients in order to isolate platelet lysate. After the application of the above treatments, patients were evaluated with an International Index of Erectile Function (IIEF-5) questionnaire, penile triplex, and reported morning erections. After MSCs and PL administration, patients presented improved erectile function after 1 and 3 months of follow-up. A statistically significant difference was observed in the IIEF-5 score before and after administration of both treatments after the first month ($p < 0.05$) and the third month ($p < 0.05$). No statistically significant difference was observed in the IIEF-5 score between group A and B patients. All patients were characterized by improved penile triplex and increased morning erections. No severe adverse reactions were observed in any patient except a minor pain at the site of injection, which was in the limits of tolerability. The results of this study indicated the satisfactory use of MSCs and PL in ED. MSCs in combination with PL or PL alone seems to be very promising, especially without having the negative effects of the current therapeutic treatment.

Keywords: erectile dysfunction; MSCs; stem cells; platelet lysate; IIEF-5 questionnaire

1. Introduction

Erectile dysfunction (ED) is a common pathology in men and it is estimated that 30 million men are suffering from a different degree of this pathology [1]. It is estimated that 50% of men ages 40 to 70 years will develop ED in the near future [1,2]. ED is the inability to attain or maintain satisfactory penile erection for sexual intercourse. In this way, ED affects a man's quality of life, as well as his partners' [3].

The corpora cavernosa plays a significant role in establishing an erection [4]. Corpora cavernosa consists of a lattice of sinusoids, which are covered by a single layer of endothelial cells (ECs), multiple layers of circular and longitudinal oriented cavernous smooth muscle cells (CSMCs), and the cavernous nerves (CNs) [4–6]. Upon stimulation, the activated neuronal NOS (nNOS) produces NO that leads to the relaxation of the cavernosal smooth muscle cells (CSMCs), which in turn let the blood to fill up the sinusoids [4,7]. The increased intracavernosal pressure compresses the penile veins against the tunica albuginea further decreasing the blood outflow and helps achieve a full erection, which is maintained by the NO produced by the endothelial nitric oxide synthase (eNOS) [4,7]. Furthermore, ECs are expressing significant levels of NO during intercourse, which maintain CSMCs in this induced state [4,7].

Due to damage in key components of erections, such as endothelial cells (ECs), cavernous smooth muscle cells (CSMCs) and neuronal cells, ED may occur [4,8–10]. Possibly, during radical prostatectomy, cavernous nerves could be damaged, causing long-term consequences that include diminished production of NO, atrophy of CSMCs, and ECs. This atrophy also could induce penile fibrosis and ECs and CSMCs apoptosis, resulting to the development of penile fibrosis [4]. ED may also occur in men diagnosed with diabetes melitus (DM), a chronic disease that affects more than 371 million people worldwide. DM may also impact and damage both the macrovascular and microvascular systems. Further, it is estimated that men diagnosed with DM have a threefold increased risk for ED [11].

Currently, ED treatments include the use of various pharmaceutical agents. The most widely used pharmaceutical agents are phosphodiesterase type-5 inhibitors (PDE5-I) [7]. Although, in some patients a single dose of PDE5-I might obtain or maintain a successful erection upon a sexual stimulus, some patients need to have repetitive doses of these agents in order to acquire or maintain a successful erection. However, if the dosage is not correct in the aforementioned agents, they can be accompanied by adverse reactions. Although some patients are having complicated health issues such as cardiovascular disease (CVD) and DM, their suitability can be reduced significantly [4,7]. In addition, the use of the pharmaceutical agents cannot be considered as curative, since the patients need to use them before sexual intercourse. Under this scope, alternative strategies must be found in order to manage properly or even treat ED. For this purpose, mesenchymal stromal cells (MSCs) could be good candidates for the treatment of ED [3].

MSCs can be isolated by several sources of human body, including bone marrow (BM), adipose tissue (AT), Wharton's Jelly (WJ) tissue, umbilical cord blood (UCB), and neonatal teeth [12–14]. MSCs are known for their immunoregulatory and immunosuppressive functions and have been administrated in patients with autoimmune disorders such as multiple sclerosis (MS), amyotrophic lateral sclerosis (ALS), and Chron's disease. During the last century, a great effort has been performed to establish their regenerative properties, by using them in tissue engineering and regenerative medicine approaches [12–14]. In addition, MSCs have the capability of multipotential differentiation to other lineages such as "osteogenic", "adipogenic", and "chondrogenic". Moreover, several research groups indicate the possible differentiation of MSCs towards "neuronal" cell lineages [12–16].

Recently, injections of autologous PL are used in regenerative medicine approaches with promising results. PL contains a significant number of growth factors such as platelet derived growth factor (PDGF), transforming growth factor—β11 (TGF-β11), vascular endothelial growth factor (VEGF), epidermal growth factor (EGF), platelet derived angiogenesis factor (PDAF), and insulin like growth factor (IGF), which is derived from platelets [15,16]. The clinical efficacy of PL depends on the concentration of growth factors that may act as transmitters, inducing wound healing and regeneration of damaged tissues [17,18].

The aim of this study was to define and quantify any improvement in erectile function of ED patients injected with ADMSCs suspended in PL or PL only. Furthermore, the collected data could be used to evaluate the feasibility of the treatment, to define any potential side effects and to estimate the sample size that is needed to design properly the next step, which is the performance of clinical trial.

2. Materials and Methods

2.1. Study Design

This study is a prospective phase 1, single center pilot study, which has been approved by the Scientific Committee of Attikon Hospital (Ref No 006), Athens, Greece. All patients were informed of the study design and signed a written informed consent in accordance to Helsinki declaration.

The patients were divided into two groups. Group A involved patients ($n = 5$) who received adipose derived MSCs (adipose derived mesenchymal stem cells (ADMSC)) with PL and group B involved patients ($n = 3$) who received only PL (Figure 1). Inclusion criteria were organic ED due to diabetes mellitus, hypertension, hypercholesterolaemia, and Peyronie disease. Detailed descriptions of each patient characteristics are listed in Table S1. Hormonal and metabolic evaluation were performed in all patients and included testosterone, estradiol, LH, FSH, PRL, FT3, FT4, TSH, α-FP, CEA, CA 19-9, glucose, cholesterol, triglycerides, and PSA (Table S2). In addition, CT scans of the abdomen, thorax, and brain was performed in all patients in order to exclude other pathologies. Evaluation of ED was performed by penile triplex with intracavernosal injection (ICI) of vasodilators and a thorough IIEF-5 questionnaire.

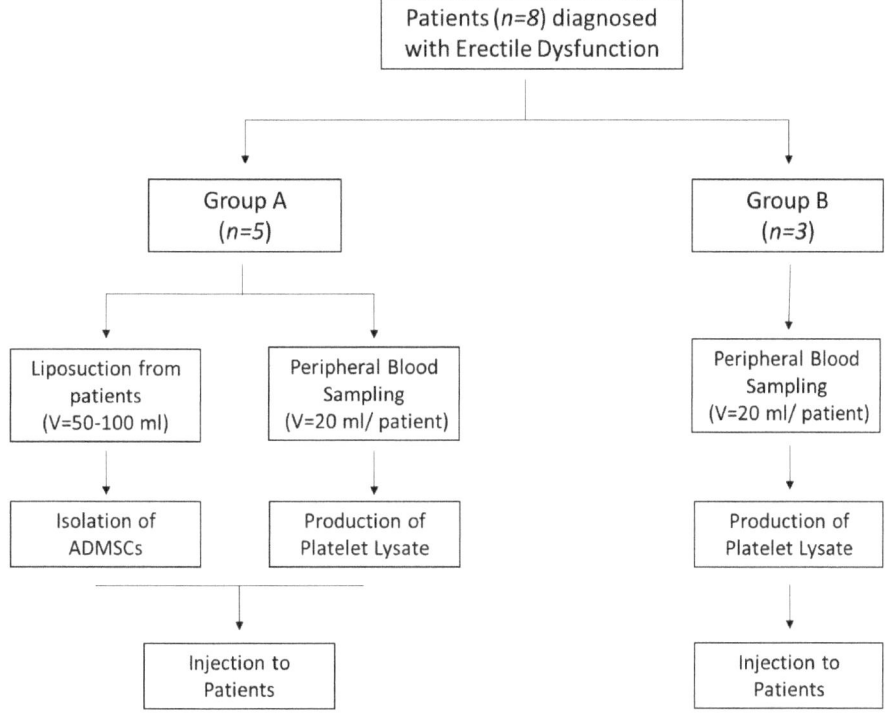

Figure 1. Study overview.

Specifically, penile triplex was performed using a Doppler ultrasonography device (Chison Qbit7 Ampronix, Medical Imaging Software, Milpitas, CA, USA). Ultrasonograms were obtained in all patients before and after the administration of ADMSCs with PL or PL monotherapy. Each patient was injected with 20 μg of alprostadil (Pfitzer, New York, NY, USA) to cause vasodilation in penile vasculature, thus resulting in an erection. Peak Systolic Velocity (PSV) and End Diastolic Velocity (EDV) were measured every 5 min for a total period of 20 min. PSV and EDV both were measured in cm/s.

Exclusion criteria for patients were lack of sexual interest, neurologic or hormonal ED, penis injuries others than Peyronie's disease, and all cases of cancer. Moreover, all patients enrolled in this pilot study were instructed not to stop or change the medication they used for ED during the follow-up period. Each patient's medication for ED is listed in Table S3. According to the findings, during the evaluation period, their medication was properly adjusted.

2.2. Isolation and Expansion of ADMSCs

Lipoaspiration was performed from all patients of group A in order to isolate ADMSCs. Isolation of ADMSCs was performed in compliance with Good Manufacturing Practices (GMPs) at clean rooms provided by the Hellenic Cord Blood Bank (HCBB) of Biomedical Research Foundation Academy of Athens (BRFAA). Briefly, adipose tissue from lipoaspiration was extensively washed in Phosphate Buffer Saline 1× (PBS 1×, Gibco, Life Technologies, Grand Island, NY, USA) for blood removal. Then, the supernatant was removed and enzymatically treated with equal volume of collagenase 1 mg/mL (Sigma-Aldrich, Darmstadt, Germany) at 37 °C in orbital shaker for a maximum of 3 h. Inactivation of collagenase was performed with the addition of PBS 1×, followed by centrifugation at 500 g for 6 min. The supernatant was discarded, the pellet was resuspended in complete cell culture medium, and transferred to 25 cm^2 cell culture flasks (Costar, Corning Life, Canton, MA, USA) in humidified atmosphere.

After 10 days of incubation, the cell cultures were microscopically checked and upon reaching 70–80% confluency, the ADMSCs were detached with 0.25% trypsin- EDTA solution (Sigma-Aldrich, Darmstadt, Germany), washed with PBS 1×, and replated to 75 cm^2 flasks (Costar, Corning Life, Canton, MA, USA). The same procedure was repeated until the cells reached passage 4. The medium of cell cultures was changed biweekly. Complete culture medium consisted of α-Minimum Essentials Medium (α-MEM, Gibco, Life Technologies, Grand Island, NY, USA) supplemented with 20% v/v Fetal Bovine Serum (FBS, Gibco, Life Technologies, Grand Island, NY, USA) 1% v/v Penicillin/ Streptomycin (Gibco, Life Technologies, Grand Island, NY, USA), and 1% L-glutamine (Gibco, Life Technologies, Grand Island, NY, USA).

2.3. Differentiation Potential of ADMSCs

ADMSCs were differentiated to "osteocytes", "adipocytes", and "chondrocytes" in order to establish their multilineage differentiation potential.

ADMSCs ($n = 5$) were differentiated to "osteocytes" using the StemProTM Osteogenic Differentiation kit (Cat. No A1007201, Gibco, ThermoFischer, Waltham, MA, USA) according to manufacturer's instructions. Finally, Alizarin red S (Sigma-Aldrich, Darmstadt, Germany) staining was used in order to validate the successful differentiation of ADMSCs to "osteocytes".

In addition, ADMSCs ($n = 5$) were differentiated to "adipocytes" with the use of StemProTM Adipogenic Differentiation kit (Ca. No. A1007001, Gibco, ThermoFischer, Waltham, MA, USA) according to manufacturer's instructions. The successful differentiation of ADMSCs into "adipocytes" was established by Oil-Red O (Sigma-Aldrich, Darmstadt, Germany) staining. Furthermore, "chondrogenic" differentiation of ADMSCs ($n = 5$) was performed using StemProTM Chondrogenesis Differentiation Kit (Cat. No A1007101, Gibco, ThermoFischer, Waltham, MA, USA). Finally, Alcian blue staining was performed in order to establish the successful "chondrogenic" differentiation of ADMSCs.

2.4. Growth Kinetics and Cell Viability of ADMSCs

Growth kinetics, including total cell number, cell doubling time (CDT), population doubling (PD), and cell viability were determined in ADMSCs after each passage until passage 5. For this purpose, 2×10^5 ADMSCs ($n = 5$) were seeded in 75 cm^2 flasks (Costar, Corning Life, Canton, MA,

USA). The total cell number of ADMSCs was counted with the use of Neubauer slide (Celeromics, Valencia, Spain). The estimation of cell viability after each passage was performed, using Trypan blue (Sigma Aldrich, St Louis, MO, USA). The determination of total cell number and cell viability of ADMSCs were performed by two different observers. CDT and PD were estimated according to the following Equations:

$$CDT = \frac{\log_{10}(N/N0)}{\log_{10}(2)} \times (T) \tag{1}$$

and

$$PD = \frac{\log_{10}(N/N0)}{\log_{10}(2)} \tag{2}$$

where N was the number of cells at the end of the culture, $N0$ was the number of seeded cells, and T was the culture duration in hours.

2.5. Immunophenotypic Analysis of ADMSCs

The immunophenotypic analysis of ADMSCs was performed according to the following panel of monoclonal antibodies. Specifically, ADMSCs ($n = 3$) at passage 4 were analyzed for CD90, CD105, CD73, CD29, CD19, CD31, CD45, CD34, CD14, CD3, CD19, HLA-DR and HLA-ABC. The CD90, HLA-ABC, CD29, CD19, CD31, and CD45 were fluorescein isothiocyanate (FITC) conjugated, while CD105, CD73, CD44, CD3, CD34, CD14, and HLA-DR were phycoerythrin conjugated. All monoclonal antibodies were purchased from Immunotech (Immunotech, Beckman Coulter, Marseille, France). The immunophenotypic analysis was performed in Cytomics FC 500 flow cytometer coupled with CXP Analysis software (Beckman Coulter, Marseille, France).

2.6. Preparation of Platelet Lysate

A total of 20 mL of peripheral blood was received from patients of group A ($n = 5$) and group B ($n = 3$) in order to isolate platelet rich plasma (PRP). The blood sampling was performed in 5 mL citrate phosphate dextrose adenine (CPDA) treated vacutainer tubes. Then, centrifugation was performed at 160 g for 20 min, followed by the isolation of plasma layer which contained the platelets (PLTs). A second centrifugation was performed at 420 g for 15 min. Finally, the volume of PRP was reduced in order to obtain the desired platelet number. PRP was stored at $-80\,^\circ$C for 48 h. Upon use, PRP was thawed, forming the PL, which was finally injected to the patients.

2.7. ADMSCs and PL Administration

ADMSCs were trypsinized, centrifuged, and the cell pellet was resuspended in 2 mL of PL in group A. Patients of group B were injected only with PL. Prior to injection, the base of penis was clamped and remained for a time period of 10 min. ADMSCs resuspended in PL or PL only were injected directly in corpora cavernosum of penis. After the injection, the patients were reevaluated on the first and third of the month. Patients follow up included physical and andrological evaluation, IIEF-5 questionnaire, and penis triplex.

2.8. Statistical Analysis

In this initial pilot study, no sample size calculations were conducted. Statistical analysis was performed with Graph Pad Prism v 6.01 (GraphPaD Software, San Diego, CA, USA). IIEF-5 scores were analyzed by Friedman's test for multiple non-parametric comparisons and then Mann-Whitney test was also applied. Statistically significant difference between group values was considered when p value was less than 0.05. Indicated values were presented as mean $\pm$ standard deviation.

3. Results

3.1. Characterization of ADMSCs

ADMSCs were characterized by fibroblastic morphology, which was retained until passage 4 (Figure 2). In addition, ADMSCs were successfully differentiated into "osteocytes", "adipocytes", and "chondrocytes". Specifically, ADMSCS differentiated from "osteocytes" and exhibited calcium deposits, which were visible by the Alizarin Red S staining (Figure 3A). Furthermore, the produced lipid droplets of the differentiated "adipocytes" were stained successfully by the Oil Red O staining (Figure 3A). Finally, the ADMSCs were capable to be differentiated into "chondrocytes" as indicated by the histological stains Toluidine and Alcian blue (Figure 3A).

The average cell number of ADMSCs at passage 4 was 47×10^6 (Figure 3B). The average CDT and PD of ADMSCs at passage 4 was 45 ± 6 h and 12 ± 0.46, respectively (Figure 3D,E).

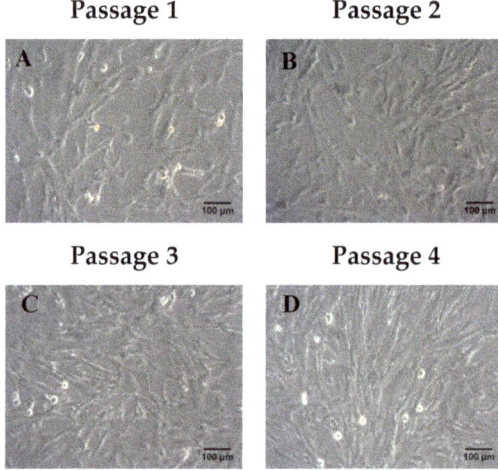

Figure 2. Morphological characteristics of adipose derived mesenchymal stem cells (ADMSC) until they reached passage 4. Representable images of ADMSCs from passage 1–4 (**A–D**). Scale bars 100 µm.

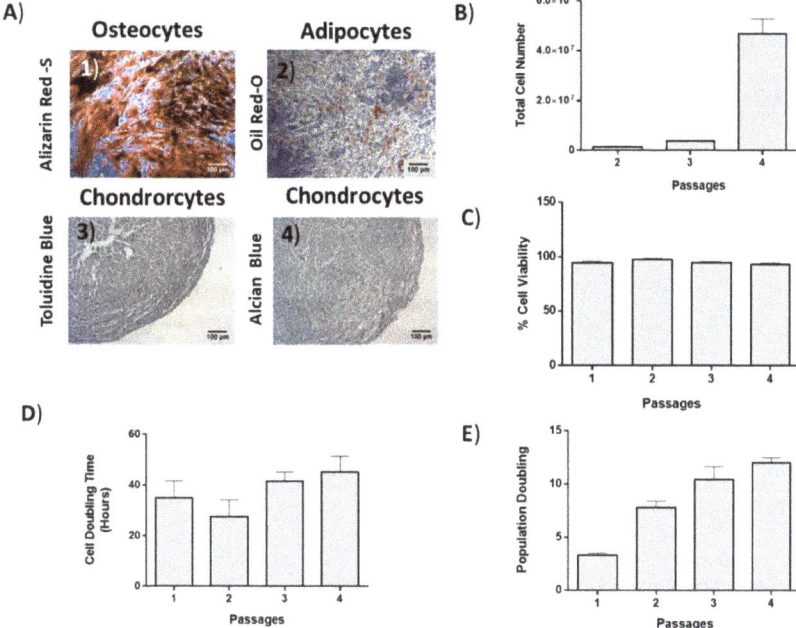

Figure 3. Differentiation potential, growth kinetics and cell viability of ADMSCs. (**A**) ADMSCs were successfully differentiated to "osteocytes", "adipocytes", and "chondrocytes". Scale bars 100 μm. (**B**) Total cell number, (**C**) cell viability, (**D**) cell doubling time (CDT), and (**E**) population doubling (PD) of ADMSCs.

3.2. Immunophenotypic Analysis

Flow cytometric analysis showed the positive expression of ADMSCs were positive (over 95%) for CD73, CD90, CD105, HLA-ABC, CD29, and CD44 and negative (below 2%) for CD3, CD19, CD31, HLA-DR, CD34, and CD45 (Figure 4).

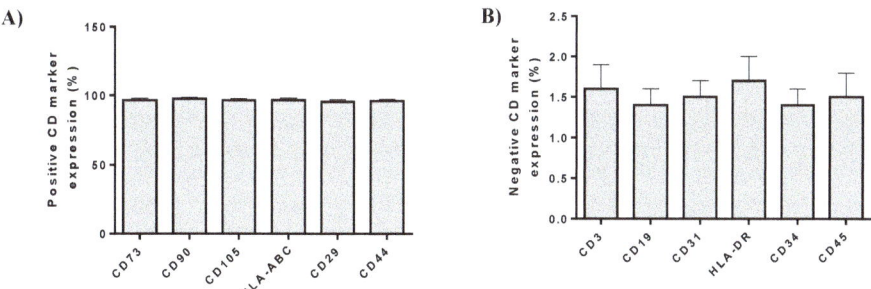

Figure 4. (**A**) Immunophenotypic analysis of ADMSCs by flow cytometer. Positive expression (%) of cell doubling (CD) markers in ADMSCs. (**B**) Negative expression (%) of CD markers in ADMSCs.

3.3. Production of PL

PL was successfully isolated from all patients of the current study. Specifically, a total volume of 2.2 ± 0.3 mL of PL which contained $1708 \pm 76 \times 10^6$ PLTs was isolated and used in patients of group A (Figure 5). In group B, a total volume of 2.3 ± 0.4 mL PL with $1693 \pm 52 \times 10^6$ PLTs was isolated

(Figure 5). A detailed description of PL characteristics is listed in Table S4. No statistically significant difference was observed either in PL volume or total PLTs between group A and B.

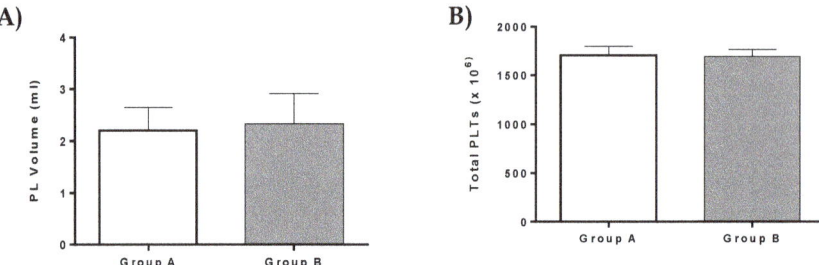

Figure 5. PL volume and total platelets (PLTs) of patients. (**A**) Platelet lysate (PL) volume of patients from group A and B. (**B**) Total number of PLTs of patients from group A and B. No statistically significant difference was observed in PL volume and total number of PLTs between group A and group B ($p > 0.9$).

3.4. Patient's Follow-up

Patients of group A received $38.9 \pm 14.4 \times 10^6$ ADMSCs in combination with 2.2 ± 0.3 mL of PL. A detailed description of ADMSCs administration to each patient is presented in table S4. Prior to ADMSCs administration, all patients were unable to have successful intercourse without the use of oral PDE5i or ICI. Following ADMSCs administration, the erectile function was improved in all patients. The majority of patients in group A were characterized by an increase in IIEF-5 score after three months of evaluation (Table 1). In addition, the patients were characterized by an improved trend on Peak Systolic Velocities (PSV), while there was a more variable pattern on the End Diastolic Velocities (EDV) as it is indicated in Table 2. Patient 1 and 2 after ADMSCs administration reported morning erections, successful sexual intercourse and their medication changed from ICI (prior to treatment) to PDE5i. Patient 3 did not present any improvement in IIEF-5 score, although he had improvement in morning erections, his medication changed from ICI (prior treatment) to PDE5i. In addition, patient 4 after the first month of ADMSCs administration noticed morning erections, while patient 5 noticed morning erections after three months. These patients achieved to maintain satisfactory unassisted erections during the whole time of sexual intercourse.

Patients of group B (Patients 6–8) received 2.3 ± 0.4 mL of PL. All patients in this group experienced an increase in PSV, EDV, and IIEF-5 score after the first and third month of evaluation. All patients reported improved erectile function and noticed morning erections. Moreover, patients were able to have unassisted erections that were maintained during sexual intercourse. Patient 6 tried several treatments before PL administration including PDE5i, ICI, and shock wave, but his erections did not improve. Following PL administration, this patient was able to have unassisted erection for intercourse in 70% of the times. Patient 7 achieved a satisfactory erection, which was maintained successfully during sexual intercourse. Patient 8 reported improved erectile function and achieved to have unassisted intercourse in 30% of the times. All patients reported improved morning erections. In addition, the pharmaceutical treatment of these patients was gradually reduced. Their erectile function was not altered, thus having successful sexual intercourse during the three months of evaluation.

There was statistically significant difference in IIEF-5 score in both groups before and after administration. Specifically, statistically significant difference was observed in IIEF-5 score before and after administration (first month, $p < 0.05$, and 3rd month, $p < 0.05$, Figure 6A,B). No statistically significant difference was observed in IIEF-5 score between patients of group A and B (Figure 6C).

Table 1. IIEF-5 scores.

	Patient	Before Administration	1st Month	3rd Month
Group A	1	6	17	12
	2	10	12	12
	3	6	5	6
	4	14	20	22
	5	16	16	20
Group B	6	12	20	20
	7	9	16	19
	8	9	16	15

Table 2. Penile triplex results.

	Patient	Before Administration (cm/s)	1st Month (cm/s)	3rd Month (cm/s)
Group A	1	35/11	30.5/7.8	39/12
	2	40.3/8.6	25.4/6.0	40.8/11.0
	3	16.1/4.7	35/8.9	61.2/20.6
	4	57/15	78.2/16.6	97.9/22
	5	45.5/17.8	49.4/14.7	62.6/25.8
Group B	6	69.9/18.7	81.9/13.4	79/12.7
	7	40/0	52.7/−10	101.5/0
	8	60/0	65.9/6.9	63.1/−7.6

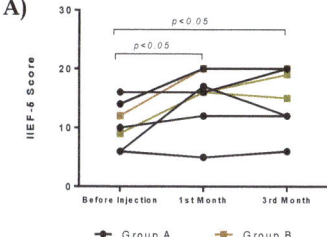
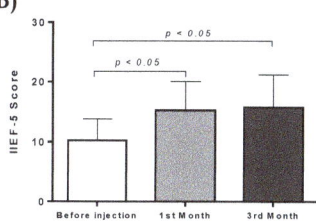
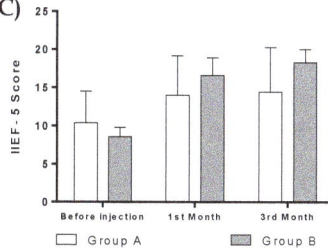

Figure 6. Effect of ADMSCs and PL therapy. IIEF scores of each patient from group A and B after intracavernous injections. (**A**) Patients improved their erectile function and had statistically significant IIEF scores after the first and third months. (**B**) IIEF score of all patients received intracavernous injections. Statistically significant difference in IIEF score of patients before the initiation of treatment and first ($p < 0.05$) and third ($p < 0.05$) month. (**C**) IIEF score from patients of group A and B.

3.5. Side Effects

No significant side effects or complications during administration of ADMSCs with PL or PL alone were reported by the patients from both groups. All patients felt a minor pain at the time of injection. The pain was more intense in group B patients and resolved spontaneously after a short time. During follow-up, no adverse reactions were reported.

4. Discussion

ED globally affects a large number of adult men and significantly reduces their quality of life. Current treatments involved the use of PDE5-I, which induces mild side effects in 70% of patients. Moreover, the use of PDE5-I is limited in patients with cardiovascular disease and diabetes melitus [4,7]. Second line treatments for ED involved the use of vacuum devices and intrecavernosal injections with PDE5-I [11]. To date, all these approaches have only temporal effects in patients suffering from ED. Recently, stem cell therapy and its derivatives have gained great attention in reversing pathological states and therefore may act as a potent ED treatment [3].

The aim of this study was to assess the safety and efficacy of the current prospective treatment in ED. In this study, autologous ADMSCs supplemented in PL and PL monotherapy were used for the treatment of ED.

Specifically, intracavernous injections of ADMSCs with PL or PL alone in patients from both groups were feasible, safe, and well-tolerated. No serious adverse events were reported after three months of patients follow-up. On physical examination, no alteration in body temperature, blood pressure, heart rate, and respiratory rate was observed.

In this study, two different groups were used. In group A, injection of ADMSCs resuspended in PL was applied, while group B involved patients that were injected only with PL. ADMSCs were characterized by fibroblastic morphology, successfully differentiated to "osteocytes", "adipocytes", and "chondrocytes", expressed CD73, CD90, and CD105 over 95%, while lacking the expression of CD34, CD45, and HLA-DR, fulfilled in this way the minimal criteria of International Society for Cellular Therapy [14]. The data indicated that ADMSCs were not characterized by different properties compared to other MSCs sources, such as Wharton's Jelly tissue, umbilical cord blood, and bone marrow. Moreover, no extended invasive surgical procedures are required for the isolation of the adipose tissue. Typically, from 1 g of adipose tissue, more than 3.5×10^5 MSCs can be isolated and used immediately or after expansion, in regenerative medicine approaches [12,13].

ADMSCs reached passage 4 successfully and were negative for aerobic, anaerobic, and mycoplasma contamination. The above data indicated that ADMSCs was a properly defined cell population, processed under GMPs conditions. In the literature, MSCs have been used successfully for treatment of various human disorders including autoimmune diseases [12,13]. In addition, properly defined MSCs secrete a variety of growth factors and cytokines which are contributing to tissue regeneration [12]. Moreover, it has been shown that despite the age discrepancy between patients, ADMSCs were successfully isolated and expanded without significant alteration in their CDT, PD, and cell viability.

This study also involved the administration of PL in patients with ED. Specifically, PL that was used in patients of group A and group B did not present any significant alteration either in the number of isolated PLTs or the injected volume. In addition, PL possesses a rich source of growth factors, which can be derived efficiently from patient's peripheral blood [16–18]. To date, in the literature, PL has been used in various regenerative medicine approaches.

After administration and three months of follow-up assessment, the majority of patients from both groups were characterized by improved IIEF-5 score, penile triplex, and a greater number of morning erections. All patients from both groups were able to have successful sexual intercourse. The medication of three patients of group A was changed from ICI to oral PDE5i. In addition, two patients from group A and all patients from group B had unassisted satisfactory erections. In these patients the pharmaceutical treatment was reduced gradually during the three months of evaluation. Despite their reduced treatment, these patients were able to perform successful sexual intercourse. Indeed, statistically significant differences in IIEF-5 score were found either in group A or B, before and after treatment (first and third month). In addition, group B presented higher IIEF-5 score in comparison to group A, although no statistically significant difference was found between these two groups. Patients in both groups experienced improvement after treatment. Although there is no statistical difference between groups, patients in group B showed improvement in their IIEF score and penile triplex results. This fact implies that PL monotherapy seemed to have better results in ED compared to ADMSCs administration. More investigation is needed to be performed in order to safely conclude which approach might have a better outcome in patients with ED.

The improvement of erectile function in patients from both groups may be as a result to the paracrine effects both of ADMSCs and PL. There is an increasing number of studies, where the MSCs have been administrated, improving the condition in various pathologies such as osteoarthritis, bone and cartilage damage, and even autoimmune disorders including Crohn's disease, multiple sclerosis, and ALS [19–22]. On the other hand, platelet lysate and its containing growth factors have been reported previously for their successful use in wound and burn healing [15,16]. In this way, both treatments may rescue endothelial dysfunction, which is common in ED, elevating the production of NO, thus improving the overall erectile function [4,8–10]. In our study, due to an underlying disease of the patients such as diabetes, hypertension, and hypercholesterolaemia, the endothelial dysfunction might be a possible explanation, and this might explain the improvement of post treatment. However, more research must be performed towards this way, in order to obtain safe results regarding the underlying therapeutic mechanism.

The results of our study seemed to be in accordance with previous published reports [23–28]. In the current study, the safety and efficacy of ADMSCs and PL injection were assessed. On the other hand, in the study of Bahk et al. [26], where single intracavernous injections of allogeneic umbilical cord blood stem cells were performed, no safety regarding the intracavernous injections was assessed. Moreover, in the current study, the patients did not perform any radical prostatectomy, but were characterized by other disorders such as hypertension and high blood glucose levels. However, the administration of ADMSCs resuspended in PL or the administration with PL only seemed to have a positive effect in erectile function, as has been reported by others [23,24,27,28].

This study was also characterized by several limitations in its performance. This pilot study was unblinded lacking the control group. In addition, group B involved only three patients. In order to obtain safe conclusions regarding the regenerative potential of ADMSCs and PL, further evaluation must be performed including follow up assessment after 6 and 12 months.

5. Conclusions

The results of the current study were very promising regarding the function improvement of ED patients. ADMSCs resuspended in PL or only PL injections could positively contribute to the treatment of ED. After three months of follow-up, patients injected only with PL seemed to have comparable outcome to patients that were treated with ADMSCs resuspended in PL. Further evaluation must be performed in order to safely conclude which approach might have the best outcome in patients with ED.

The future goal of this study is to enroll a higher number of patients who will be evaluated for their erectile function over a longer time period of time. ED compromise a wide socioeconomic burden, affecting a great number of men, and any possible therapeutic strategy without the adverse effects of previous treatments may be very promising.

Supplementary Materials: The following are available online at http://www.mdpi.com/2306-5354/6/1/21/s1, Table S1. Patient's age and comorbidities. Table S2. Hormonal and metabolic evaluation of all patients. Table S3. Patients' medication before the administration of ADMSCs with PL or PL. Table S4. Number of administrated ADMSCs to each patient.

Author Contributions: Conceptualization, V.P.; Data curation, P.M.; Investigation, C.L., M.C. and C.D.; Methodology, V.P., E.M., P.M., I.G, Z.D. and N.K.; Project administration, V.P.; Supervision, V.P.; Writing—original draft, P.M.; Writing—review & editing, E.M. and C.S.-G. All authors contributed for the successful accomplishment of this work.

Funding: This research received no funding.

Conflicts of Interest: The authors declare no conflict of interest.

References

1. Nehra, A. Oral and non-oral combination therapy for erectile dysfunction. *Rev. Urol.* **2007**, *9*, 99–105. [PubMed]
2. Mangır, N.; Türkeri, L. Stem cell therapies in post-prostatectomy erectile dysfunction: A critical review. *Can. J. Urol.* **2017**, *24*, 8609–8619. [PubMed]
3. Mangir, N.; Akbal, C.; Tarcan, T.; Simsek, F.; Turkeri, L. Mesenchymal stem cell therapy in treatment of erectile dysfunction: Autologous or allogeneic cell sources? *Int. J. Urol.* **2014**, *21*, 1280–1285. [CrossRef] [PubMed]
4. Reed-Maldonado, A.B.; Lue, T.F. The Current Status of Stem-Cell Therapy in Erectile Dysfunction: A Review. *World J. Men's Health* **2016**, *34*, 155–164. [CrossRef] [PubMed]
5. Lue, T.F. Erectile dysfunction. *N. Engl. J. Med.* **2000**, *342*, 1802–1813. [CrossRef] [PubMed]
6. Prieto, D. Physiological regulation of penile arteries and veins. *Int. J. Impot. Res.* **2008**, *20*, 17–29. [CrossRef] [PubMed]
7. Lin, C.S.; Xin, Z.C.; Wang, Z.; Deng, C.; Huang, Y.C.; Lin, G.; Lue, T. Stem cell therapy for erectile dysfunction: A critical review. *Stem Cells Dev.* **2012**, *21*, 343–351. [CrossRef] [PubMed]
8. Mulhall, J.P.; Bella, A.J.; Briganti, A.; McCullough, A.; Brock, G. Erectile Function Rehabilitation in the Radical Prostatectomy Patient. *J. Sex. Med.* **2010**, *7*, 1687–1698. [CrossRef] [PubMed]
9. Iacono, F.; Giannella, R.; Somma, P.; Manno, G.; Fusco, F.; Mirone, V. Histological alterations in cavernous tissue after radical prostatectomy. *J. Urol.* **2005**, *173*, 1673–1676. [CrossRef] [PubMed]
10. Fode, M.; Ohl, D.A.; Ralph, D.; Sønksen, J. Penile rehabilitation after radical prostatectomy: What the evidence really says. *BJU Int.* **2013**, *112*, 998–1008. [CrossRef] [PubMed]
11. Dashwood, M.R.; Crump, A.; Shi-Wen, X.; Loesch, A. Identification of neuronal nitric oxide synthase (nNOS) in human penis: A potential role of reduced neuronally-derived nitric oxide in erectile dysfunction. *Curr. Pharm. Biotechnol.* **2011**, *12*, 1316–1321. [CrossRef] [PubMed]
12. Chatzistamatiou, T.K.; Papassavas, A.C.; Michalopoulos, E.; Gamaloutsos, C.; Mallis, P.; Gontika, I.; Panagouli, E.; Koussoulakos, S.L.; Stavropoulos-Giokas, C. Optimizing isolation culture and freezing methods to preserve Wharton's jelly's mesenchymal stem cell (MSC) properties: An MSC banking protocol validation for the Hellenic Cord Blood Bank. *Transfusion* **2014**, *54*, 3108–3120. [CrossRef] [PubMed]
13. Mallis, P.; Boulari, D.; Michalopoulos, E.; Dinou, A.; Spyropoulou-Vlachou, M.; Stavropoulos-Giokas, C. Evaluation of HLA-G Expression in Multipotent Mesenchymal Stromal Cells Derived from Vitrified Wharton's Jelly Tissue. *Bioengineering* **2018**, *5*, 95. [CrossRef] [PubMed]
14. Dominici, M.; Le Blanc, K.; Mueller, I.; Slaper-Cortenbach, I.; Marini, F.C.; Krause, D.S.; Deans, R.J.; Keating, A.; Prockop, D.J.; Horwitz, E.M. Minimal criteria for defining multipotent mesenchymal stromal cells. The international society for cellular therapy position statement. *Cytotherapy* **2006**, *8*, 315–317. [CrossRef] [PubMed]
15. Lubkowska, A.; Dolegowska, B.; Banfi, G. Growth factor content in PRP and their applicability in medicine. *J. Biol. Regul. Homeost. Agents* **2012**, *26*, 3S–22S. [PubMed]
16. Christou, I.; Mallis, P.; Michalopoulos, E.; Chatzistamatiou, T.; Mermelekas, G.; Zoidakis, J.; Vlahou, A.; Stavropoulos-Giokas, C. Evaluation of Peripheral Blood and Cord Blood Platelet Lysates in Isolation and Expansion of Multipotent Mesenchymal Stromal Cells. *Bioengineering* **2018**, *5*, 19. [CrossRef] [PubMed]
17. Schallmoser, K.; Strunk, D. Preparation of pooled human platelet lysate (pHPL) as an efficient supplement for animal serum-free human stem cell cultures. *J. Vis. Exp.* **2009**, *32*, 1523. [CrossRef] [PubMed]

18. Fekete, N.; Gadelorge, M.; Fürst, D.; Maurer, C.; Dausend, J.; Fleury-Cappellesso, S.; Mailänder, V.; Lotfi, R.; Ignatius, A.; Sensebé, L.; et al. Platelet lysate from whole blood-derived pooled platelet concentrates and apheresis-derived platelet concentrates for the isolation and expansion of human bone marrow mesenchymal stromal cells: Production process, content and identification of active components. *Cytotherapy* **2011**, *14*, 540–554.
19. Xia, T.; Yu, F.; Zhang, K.; Wu, Z.; Shi, D.; Teng, H.; Shen, J.; Yang, X.; Jiang, Q. The effectiveness of allogeneic mesenchymal stem cells therapy for knee osteoarthritis in pigs. *Ann. Transl. Med.* **2018**, *6*, 404. [CrossRef] [PubMed]
20. Southworth, T.M.; Naveen, N.B.; Tauro, T.M.; Leong, N.L.; Cole, B.J. The Use of Platelet-Rich Plasma in Symptomatic Knee Osteoarthritis. *J. Knee Surg.* **2018**, *13*, 37–45. [CrossRef] [PubMed]
21. Petrou, P.; Gothelf, Y.; Argov, Z.; Gotkine, M.; Levy, Y.S.; Kassis, I.; Vaknin-Dembinsky, A.; Ben-Hur, T.; Offen, D.; Abramsky, O.; et al. Safety and Clinical Effects of Mesenchymal Stem Cells Secreting Neurotrophic Factor Transplantation in Patients with Amyotrophic Lateral Sclerosis: Results of Phase 1/2 and 2a Clinical Trials. *JAMA Neurol.* **2016**, *73*, 337–344. [CrossRef] [PubMed]
22. Zhang, X.M.; Zhang, Y.J.; Wang, W.; Wei, Y.Q.; Deng, H.X. Mesenchymal Stem Cells to Treat Crohn's Disease with Fistula. *Hum. Gene Ther.* **2017**, *28*, 534–540. [CrossRef] [PubMed]
23. Yiou, R.; Hamidou, L.; Birebent, B.; Bitari, D.; Le Corvoisier, P.; Contremoulins, I.; Rodriguez, A.M.; Augustin, D. Intracavernous Injections of Bone Marrow Mononucleated Cells for Postradical Prostatectomy Erectile Dysfunction: Final Results of the INSTIN Clinical Trial. *Eur. Urol. Focus* **2017**, *3*, 643–645. [CrossRef] [PubMed]
24. Yiou, R. Stem-cell therapy for erectile dysfunction. *Biomed. Mater. Eng.* **2017**, *28*, S81–S85. [CrossRef] [PubMed]
25. Yiou, R.; Hamidou, L.; Birebent, B.; Bitari, D.; Lecorvoisier, P.; Contremoulins, I.; Khodari, M.; Rodriguez, A.M.; Augustin, D.; Roudot-Thoraval, F.; et al. Rouard Safety of Intracavernous Bone Marrow-Mononuclear Cells for Postradical Prostatectomy Erectile Dysfunction: An Open Dose-Escalation Pilot Study. *Eur Urol.* **2016**, *69*, 988–991. [CrossRef] [PubMed]
26. Bahk, J.Y.; Jung, J.H.; Han, H.; Min, S.K.; Lee, Y.S. Treatment of diabetic impotence with umbilical cord blood stem cell intracavernosal transplant: Preliminary report of 7 cases. *Exp. Clin. Transplant.* **2010**, *8*, 150–160. [PubMed]
27. Haahr, M.K.; Jensen, C.H.; Toyserkani, N.M.; Andersen, D.C.; Damkier, P.; Sørensen, J.A.; Lund, L.; Sheikh, S.P. Safety and Potential Effect of a Single Intracavernous Injection of Autologous Adipose-Derived Regenerative Cells in Patients with Erectile Dysfunction Following Radical Prostatectomy: An Open-Label Phase I Clinical Trial. *EBioMedicine* **2016**, *5*, 204–210. [CrossRef] [PubMed]
28. Haahr, M.K.; Jensen, H.C.; Toyserkani, N.M.; Andersen, D.C.; Damkier, P.; Sørensen, J.A.; Sheikh, S.P.; Lund, L. A 12-Month Follow-up after a Single Intracavernous Injection of Autologous Adipose-Derived Regenerative Cells in Patients with Erectile Dysfunction Following Radical Prostatectomy: An Open-Label Phase I Clinical Trial. *Urology* **2018**, *121*, 203.e6–203.e13. [CrossRef] [PubMed]

© 2019 by the authors. Licensee MDPI, Basel, Switzerland. This article is an open access article distributed under the terms and conditions of the Creative Commons Attribution (CC BY) license (http://creativecommons.org/licenses/by/4.0/).

Opinion

Introducing the Language of "Relativity" for New Scaffold Categorization

Haobo Yuan

School of Engineering, University of South Australia; Mawson Lakes Blvd, Salisbury 5095, Australia; Haobo.yuan@mymail.unisa.edu.au; Tel.: +61-08-8302-6611

Received: 10 January 2019; Accepted: 22 February 2019; Published: 26 February 2019

Abstract: Research related with scaffold engineering tends to be cross-domain and miscellaneous. Several realms may need to be focused simultaneously, including biomedicine for cell culture and 3D scaffold, physics for dynamics, manufacturing for technologies like 3D printing, chemistry for material composition, as well as architecture for scaffold's geometric control. As a result, researchers with different backgrounds sometimes could have different understanding towards the product described as 'Scaffold'. After reviewing the literature, numerous studies termed their developed scaffold as 'novel', compared with scaffolds previously designed by others using comparing criterion like 'research time', 'manufacturing method', 'geometry', and so on. While it may have been convenient a decade ago to, for example, categorize scaffold with 'Dualistic Thinking' logic into 'simple-complicated' or 'traditional-novel', this method for categorizing 'novelty' and distinguishing scaffold is insufficiently persuasive and precise when it comes to modern or future scaffold. From this departure of philosophical language, namely the language of 'relativity', it is important to distinguish between different scaffolds. Other than attempting to avoid ambiguity in perceiving scaffold, this language also provides clarity regarding the 'evolution stage' where the focused scaffolds currently stand, where they have been developed, and where in future they could possibly evolve.

Keywords: language of relativity; scaffold categorization; evolution of scaffold; seven-folder logics; cell culture; 3D scaffold; dynamicity and dimensionality; traditional scaffold; novel scaffold; future scaffold engineering; laws of system evolution; 3DPVS; vibrating nature of universe.

1. Introduction

Cell culture scaffold is defined as a class of artificially-created biomedical products used for culturing cells in vitro, through mimicking some real tissue properties. Scaffold engineering, in this connection, has developed in two chief avenues. One can be 'static into dynamic', with proven effects that dynamic cultures have benefits over static ones. In this direction, scaffolds were focused on dimensional difference, so to speak, '2D into 3D'. In addition, artificially created 3D scaffolds, which have the nature of being passive (i.e., 'static'), have been utilized, helping external culturing mimic real tissue in 3D environments with better performance, compared with traditional 2D cell culturing methods. In another direction, attempts have been made to develop cell culturing with some dynamic properties or possibly combine scaffold with some mechanical means like shakers, both of which were aimed to potentially approximate part of dynamic environments in real tissue. Provided that the new categorization of scaffold, which uses the 'evolution' criteria as the 'center of gravity', can be studied in two ways, namely the evolutional 'points' in dimensionality as well as in dynamicity. A 'point' could represent a certain combination of properties, which is organized in a definite time and place, and fulfills a definite function in one system or another. Take scaffold for instance: a 'point of the scaffold' can be designated through the number of central or predominating properties inside the scaffold. The philosophy will be more easily perceived when it comes to the seven-layer classifying process.

On other hand, this paper aims to introduce the language of 'relativity' and apply it as the novel scaffold categorization method. Currently, existing scaffolds in biomedical worlds will be defined as the scaffold No. 1, 2, 3, and 4. In this connection, each scaffold will be introduced and studied briefly. Then, the contents of each will be organized in 'summary' format, instead of a detailed 'review' or explanation. Future scaffold No. 5, which consists a scaffold of 'self-integration', will be introduced, with potential 3DPVS (3D Printed Vibratory Scaffold) as one essential part which was justified via studying scaffold No. 1, 2, and 3, as well as simultaneously investigating 'Laws of System Evolution (LSE)' into scaffold engineering. It is worth noting that the scaffold No. 5 concept includes a multiple-aspect justification and has been conducted previously by author's research group and transmitted into publications [1,2]. Scaffold No. 6 and 7 are scaffolds that might potentially appear in relatively long-term future, after the mature stage of scaffold No. 5, which will be briefly discussed later. Scaffold No. 1 to No. 7 will compose the full seven-layer evolutional ladder of scaffold engineering. Besides the categorizations, a possible further sub-classifying, possibly existing under each of the seven hierarchies, will also be briefly mentioned, which might be interesting to follow-up with in the future.

2. Introducing the Language of 'Relativity'

'Relativity' is a complicated system that has been applied in both philosophy and modern physics [3]. To understand 'relativity', we use the terms 'novel' and 'traditional' regarding scaffold as examples. From the current review of literature, the term '3D scaffold' predominantly refers to 3D passive or static scaffold. 'Advanced 3D scaffold' or 'novel 3D scaffold' refers to a number of 3D scaffolds fabricated with more sophisticated material complex and higher controlled geometries. However, these 'traditional 3D scaffolds' can be and have been termed as 'novel, advanced, or innovative' by researchers when compared with earlier 2D culture plates, platforms, substrates, or scaffolds, which literally exist as more traditional. Therefore, the term 'novel' can only be established based on 'relativity'. Misunderstanding easily occurs when using a term like 'novel cell culture scaffold' without a permanent reference object or centre and this indeed becomes a limitation for researches to communicate 'novelty' inside scaffold.

As philosophy states, exact language is necessary for exact understanding [3]. In this paper, we made efforts to define scaffold based on 'relativity' that could make it possible to know what is being discussed, from which point of view, and in what connection. The fundamental property of the new definition is that all concepts of 'scaffolds', as well as their mutual relationships, are concentrated around one idea. This idea is about 'evolution', namely the cell culture scaffold's evolution. Seven levels of scaffolds have been defined to potentially illustrate the whole evolution process of scaffold engineering, which covers the past, present, and future. With this betterment in scaffold's categorization, future scaffold researchers might be benefited, howsoever small, with the increased clarity, precision, and understanding during their studies.

3. The Novel Seven-Layer Scaffold Categorization

Using the language of 'relativity', scaffolds will be categorized in a seven-layer format, which helps indicate the relative position of each scaffold or each group of scaffolds that share some common traits inside the whole scaffold's evolutional process. Previous and major current scaffolds would be defined as scaffold No. 1, 2, and 3; future 3D Vibratory Scaffold, which will be developed via the current research period, would be termed as scaffold 5, and the bridging scaffold products between scaffold No. 1, 2, 3, and 5, would be referred to as scaffold 4. Scaffold No. 7 refers to the ideal scaffold ultimately evolved from scaffold No. 5; this evolution could practically occur in the future, but is limited by existing technology in bioscience and materials, as well as the immature development of scaffold No. 5. Scaffold bridging the scaffold No. 5 and 7 can be named as scaffold No. 6. As a matter of fact, scaffold No. 6 and 7 could merely be predicted and perceived theoretically, but much work in this direction will be needed for future researchers. The seven-fold scaffold classification will be discussed in three groups: scaffold No. 1, 2 and 3; scaffold No. 4 and 5; scaffold No. 6 and 7; and a concluding comparison of the seven levels of scaffold is illustrated in Table 1.

3.1. Scaffold Number One, Two, and Three

In this section, three scaffold classes will be introduced. Scaffold No.1, 2, and 3 constitute the predominantly major part of currently existing literature on scaffold engineering.

3.1.1. Scaffold No. 1

Scaffold No. 1 is the traditional 2D static scaffolds used for 2D cell cultures. Scaffold No. 1 is the product initiating the evolution process of scaffold engineering. Invented more than 50 years, 2D scaffolds or platforms are sometimes still used by researchers due to the inexpensive and easy-fabricating properties [4,5], though many of them have been gradually replaced by 3D scaffold, which starts as scaffold No. 2. For both dimensionality and functionality, scaffolds in this category stay at the bottom of the scaffold's evolutional ladder. This scaffold product used to be considered 'novel' when comparing it with 'traditional' 2D plastic plates or containers used in much early 2D cell culturing. The 2D substrates remain as flat surfaces with dimensions in the z-axis at a hundred nanometre level where cell growth is predominantly controlled in an x-, y-direction [5], e.g., alignment of cells along shallow grooves.

3.1.2. Scaffold No. 2

Scaffold No. 2 is the early-phase 2.5D or 3D static or passive scaffolds used for 3D cell culture; these scaffolds occupy much of the early literature concerning '3D cell culture' and they are usually designed with single and simple material composition and geometries. Scaffold No. 1 into No. 2 follows the evolution law of 'lower dimensional into higher dimensional'. If it is assumed that scaffold in this category is 'novel', then the comparison becomes reasonable only when the compared object is scaffold No. 1, which stays at a relatively lower ladder or the less developed scaffold No. 2 in same ladder. Scaffold No. 2 could also be considered as the early-stage or immature products of 3D scaffold, which will be defined as scaffold No. 3. Work of scaffold No. 2 concentres around mid-20th century, where natural extracellular matrixs (ECM) or gels composed of natural or synthetic polymers were used [6,7]. Conclusion of culturing cell on flat 2D versus 3D can significantly alter cellular responses generally originated from such approaches.

3.1.3. Scaffold No. 3

Scaffold No. 3 is the currently developed or to-be-developed 3D static scaffold used for advanced cell culture applications; these scaffolds tend to have multiple material composition, multi-functional biomedical control, and complicated architectural structures. This scaffold is also the major focus of researchers in the current scaffold engineering field. When the term '3D scaffold' has been used in past years, it mostly referred to scaffold No. 3. This scaffold group chiefly follows the law of 'simple structure and composition into a complicated and complex one'. Scaffold No. 3 includes the major part of scaffolds termed as novel or advanced by recent researchers. This subconscious belief could lie on the fact that scaffold No. 3 is surely 'novel' over scaffolds No. 1 and 2, thus leading people to take it for granted that the 'novel' scaffold is equivalent as scaffold No. 3. However, such conception is not true, according to the 'relativity' philosophy discussed earlier; if researcher were to work on scaffold No. 4 and 5 as their project 'novel scaffold design', then all scaffold No. 1, 2, and 3 would be logically categorized into a 'traditional' category, which seems to be scientific and provides with better clarity and understandability to other researchers.

The appearance of scaffold No. 3 could be partly due to the invention of 3D printing as the third industrial revolution [8,9], which flourished in tissue engineering. Opposed to 3D scaffolding brings about the concept of scaffold-free TE (tissue engineering), which, based on the assembly of building 'cell sheets' and 'spheroids' blocks, provides merits over conventional scaffolds No. 1 and 2, especially regarding tissue formation efficiency. However, it cannot replace the role of 3D scaffold, which provides a biomimetic environment that delivers or controls the release of growth and differentiation factors as well as the robust structure protecting cells from possible damage via external factors [10]. Therefore, if scaffold-free is considered, it could generally rate at evolutional ladder at scaffold No. 3 or perhaps No. 4, if a more advanced dynamic or dimensional properties are embedded within.

3.2. Scaffold Number Four and Five

This section will discuss two scaffold categorizations: scaffold No. 4 and scaffold No. 5. Future scaffold research starts from here.

3.2.1. Scaffold No. 4

Scaffold No. 4 is the scaffold with either some dynamic, partially dynamic, or active to some extent properties. Current 3D scaffolds are fabricated with shape-changing materials and can be the typical products in this category. Bridging traditional scaffold No. 1, 2, and 3, and future novel scaffold No. 5, aspects regarding scaffold No. 4 would occupy a significant positionality towards developing new scaffolds. A chief feature of scaffold No. 4 is its increasingly integrated dynamic functionality inside a 3D scaffold. Compared with scaffold No. 3, material composition would be the major difference between both. New materials, for instance smart materials, which can generate some movable or changeable actions of scaffold under specific stimulus [11,12], could be utilized experimentally on scaffold engineering. Relation of scaffold No. 4 to No. 5 is similar as the relation of scaffold No. 2 to No. 3. Further, scaffold No. 4 could be considered as the immature or preparing-stage product of scaffold No 5, whereas the dynamic properties not integrated into scaffolds as an indivisible unit or the aimed dynamicity can only partly be fulfilled by scaffold. In brief, when evolutional 'points' of dynamicity is concerned, 'traditional scaffold' starts to indicate the whole categories of scaffold No. 1, 2, and 3, which has the nature of being passive or static. In this connection, scaffold No. 4 and No. 5 could be scientifically defined as 'novel or advanced'.

In terms of the currently emerged organ-printed product or bio-printing that has a similar function as scaffold, its level of ideality might be seen as a corresponding level of scaffold 4, based on the level of advance in dimensional control and in dynamic or vibratory properties. Since bio-printing is in infancy [9,13], further development may make it evolve as a No. 5 level product. An interesting evolutionary indication of vibration on a bio-printed product or scaffold is whether or not it could controllably mimic human-like cellular vibrating in definite cell culture and to what extent it can mimic. Other scaffold products, such as scaffolds with advanced bio-restorability, bio-activity, and timely-changing properties, could be positioned at scaffold No. 4 level with similar logic. Furthermore, the use of a decellularized extracellular matrix (ECM) of a tissue is another promising trend, part of which would possibly stand out as evolutionary, similar as scaffold No. 4, considering its beneficial advances in bio-dynamicity and real tissue structuring. Moreover, the advent of whole organ decellularization brings extracellular matrix scaffolds suitable for organ engineering [14–16]. Current preliminary works in this direction can be considered at the No. 4 level and future scaffolds have the potential to arrive at a No. 5 or greater.

3.2.2. Scaffold No. 5

Scaffold No. 5 is the future 3D dynamicity-integrated Scaffold that usually unifies one chief dynamic from external to internal, which is an unattainable standard for current scaffolds. In brief, scaffold No. 5 is the scaffold of 'self-integration', for it is a scaffold that has reached 'unity'. Evolved from scaffold No. 1, 2, 3, and 4, it changes its role of being static and passive into active and dynamic. Unification of separate parts—e.g., making an external dynamic device and 3D scaffold one whole unit—is the chief feature of scaffold No. 5. To be more specific, scaffolds No. 1, 2, and 3 tend to be limited in receiving vibrations through connecting culturing platforms to external vibrators or mechanical shaker vibrators [17,18], while scaffold No. 5 is the 3D vibratory scaffold capable of generating proper vibrations, which are required by specific cell cultures in an internal and integrated manner. Such an approach would mitigate several limitations in traditional dynamic cell cultures [2]. A novel future scaffold concept, 3DPVS (3D printed vibratory scaffold) [1], might be a typical representative in the scaffold No. 5 category, which could even be potentially equivalent to scaffold No. 5. This is based on the current limitations of scaffold as studied [2], in addition to the future pointing direction, evaluated from the 'laws of system evolution' [1,19,20].

Conceptual 3DPVS is generated as a byproduct of No. 5, which concerns 3D printed technology and vibration that transforms the passive role of scaffold receiving vibration into an actively generating vibration. It would adequately include merits and properties of traditional scaffold No. 2 and 3, as well as the benefits of exact vibratory functions onto cell culture. Vibration compared with other dynamic properties in cell culture has been studied as an optimal dynamicity to be applied on scaffold No. 5. Moreover, novel 3D printing (3DP) has been the technology that bridges traditional scaffold No. 1, 2, and 3 to fabricating novel 3D vibratory scaffold. In connection with scaffold No. 5, scaffold No. 4 could logically include the group of scaffolds under testing that experiment with different vibration mechanisms and different materials. Thus, No. 4 could gradually approximate yet remain distant from the goal of scaffold No. 5. In other words, the basic logic behind scaffold No. 5 is 'static into dynamic', 'passive into active', and 'separated into integrated'. 'Vibration' inside scaffold No. 5 will be much finer and subtler compared with the 'vibration', as applied in previous scaffolds. This is perhaps a result of a molecular vibrating level. Furthermore, better applications for in vitro cell studies might be another character for scaffold No. 5. As a result, it is reasonable to predict that scaffold No. 5 would play a promising role in the near future of scaffold engineering.

3.3. Scaffold Number Six and Seven

Scaffold No. 6 and No. 7 refer to scaffolds in the relativley long-term future. They might start playing roles in future engineering when the progress of scaffold No. 5 becomes solid, applicable, and mature. At the current stage, their value might predominately be considered as conceptual and theoretical.

3.3.1. Scaffold No. 6

Scaffold No. 6 is the intermediate or bridging scaffold between scaffold No. 5 and No. 7; its related role towards scaffold No. 7 is like that of scaffold No. 4 towards No. 5. At its current stage, scaffold No. 6 and 7 remains theoretical and might not appear in the near future until research regarding scaffold No. 5 has been solid and successful. In terms of potential properties, scaffold No. 6 might open the avenue of multiple external dynamic properties, where functions in cell culture could merge with the scaffold itself. Multi-dynamicity plus integration would be the typical feature for scaffold No. 6, which evolves from scaffold No. 5. No. 6 focuses on single-dynamicity that integrates into the vibration property in 3DPVS. Compared with 'self-integration' in scaffold No. 5, scaffold No. 6 possibly attains the capability for 'multipleness-integration'. Developed from 3D, the dimensionality of scaffold No. 6 might be expanded, provided that realistic 4D (4-dimensional) engineering could come into its application. Standing relatively close to scaffold No. 7, as well as following same evolutional law, scaffold No. 6 might cease to be artificial as external cell culture scaffold and becomes more bio-mimic to real organs or tissues. To conclude, scaffold No. 6 could be the future direction following the development and innovation of scaffold No. 5. A successful design and mature application would open avenues for scaffold No. 6.

On other hand, past-to-present scaffolds or bio-products witnessed three generations of biomaterials, from common and borrowed materials to engineered implants, as well as to bioengineered implants using bioengineered materials. The revolution of materials also affects the revolution of scaffolds. Although it is not clear what the following evolution of material indicates, it could very possibly contribute to the emerge of products as scaffold No. 6.

3.3.2. Scaffold No. 7

Scaffold No. 7 is the 'ideal' future scaffold, as it is highly dimensional and dynamic, and is defined as a scaffold capable of fully mimicking, replacing, or manipulating the in vivo cell culture of 3D microenvironments. Scaffold No. 7 would reach its full development that possesses 'everything' a scaffold can possess. It might close the gap of cell culture between the external and internal. One chief evolutional point of Scaffold No. 7 dwells at dynamicity, i.e., more dynamic functions or properties beyond 'vibration', which is the chief feature of scaffold No 5, will be endowed. Thus, scaffold No. 7 will probably be the ultimate 3D or higher-dimensional dynamic scaffold, as far as the currently perceivable properties of scaffold are concerned. Above which the concept of 3D scaffold might cease to exist and further evolution ladder would become unpredictable. A scaffold No. 7 might also be viewed like perceiving current mechanical robots and the future artificial-intelligence resulted human-mimic robot, the latter of which has only been shown in science fiction. In short, scaffold No. 7 might be the perfect scaffold.

Further discussions of 'vibration' and 'dimensionality', from an evolutional point of view, substances the that universe consists of vibrations and matter, or of matter in a state of vibration, or of vibrating matter. The rate of vibration is an inverse ratio to matter's density. Higher and finer vibration generally correlates to a higher consciousness of definite objects or creatures, while low-level vibrations are connected with more mechanicalness. We could understand this from the difference between living creatures and 'dead' artificial machines. Moreover, high dynamicity could mean a higher level of vibration in anobject, whether predictable or not. Therefore, to judge the relative position of an object in the evolutionary ladder, vibration property could be used as a unique but cosmic trait, given that it is not merely understood in a narrow way. Some of the current literature claims that the use of four-dimensional printing or material is misleading, since 4D is not ordinarily perceivable by human perceptual senses. What these authors termed as 4D generally means a 3D product with timely changeable properties. However, this is not 4D, which indeed means 'time' is modifiable 'back-forth' as what can be done in other three dimensions [3,21,22]. Thus, 4D scaffold remains a long-future conceptual product, let alone a higher dimensional product beyond this. To term scaffold No. 6 and 7 with possibly higher-dimensional properties could only be a theoretical indication, while not the decisive conclusion. To summarize, after the previous description regarding the seven-layered scaffold classification, Table 1 gives a brief summarization of the categorization work as communicated.

Table 1. Summarizing table of scaffold categorization via introducing the language of 'relativity'.

Time Aspects	Previously Focused Cell Culture Scaffold			Bridging Scaffold	Short-Future Scaffold	Long-Future Scaffold	
Categorization	Scaffold No. 1	Scaffold No. 2	Scaffold No. 3	Scaffold No. 4	Scaffold No. 5	Scaffold No. 6	Scaffold No. 7
Dimensionality	2D	2.5D–3D	3D	3D	3D	3D or beyond	3D or beyond
Dynamicity	Static, Passive	Static, Passive	Static, Passive	Partly Dynamic, Active	Finely dynamic, Vibratory, Active	Highly Dynamic, Vibratory, Active	Ideally Dynamic, Vibratory, Active
Chief Feature in brief	2D plate or scaffold for 2D Cell Culture; very limited cell application; mostly replaced by scaffold No. 2 and 3; passive or static; stays at the bottom of scaffold's evolutional ladder	3D Scaffold in early stage, simple characterization and single function, static or passive	Current mainstream 3D scaffold being passive or static, with Characterizations increasingly tailored on geometrics and composition	Recent scaffold partly made or designed by smart or dynamic materials; Bridging scaffold No. 1, 2, 3 to No. 5; potentially dynamic or active; inevitable stage toward achieving scaffold No. 5	3D scaffold with fully integrated vibratory functions; tailored- or self-vibratory properties; 3DPVS as one typical d; currently under conceptual development stage	Bridging scaffold No. 5 to No. 7; remains as a concept; Multiple-dynamic functions integrated inside scaffold might appear; much sophisticated properties	Ideal Scaffold in self-perception; Fully, ideally controllable vibratility; probably close the gap between in vitro and vivo
Evolutional Ladder	Relatively Lowest	Low	Moderate	Slightly High	High	Very High	Relatively Highest

3.4. Discussion on Futher Sub-Classification of Scaffold Under the Seven-Fold

In previous sections, the seven-layer classification of scaffold has been brought up. It is logical to question the further sub-classes inside each of these seven scaffolds when scaffold is identified into one of them. A further classification inside each layer may help researchers better understand scaffolds. Work such as this will be meaningful when the seven-layer scaffold classification has become more mature and accepted. However, the basic point will still revolve around the idea of 'evolution', whereas 'relativity' will be based on 'traits' that can help scaffolds identical from one another.

In an ordinary sense, scaffold classification is done according to external traits that can be partial or unidentical from the different experience or respective researcher point of view. In further developed engineering, which we could temporarily term it as 'exact' engineering, we assume that classification can be made based on 'cosmic' traits. As a matter of fact, there could exist some 'exact' traits which can be identical for every scaffold inside each of the seven categories, allowing future researchers to establish the finer sub-class of a given scaffold with possibly the utmost exactitude, both in relation to other scaffolds as well as to its own position in the ladder of scaffold's evolutional process. From a current understanding, we suppose the 'cosmic' level of being of every scaffold could be determined via four aspects: first, by what this scaffold is made, produced, or fabricated; second, by what the scaffold 'breathes', i.e., how scaffold transmits air, nutrients, or other materials between vitro and vivo at an in-out process; third, by the cell culturing medium where the scaffold 'lives'; and fourth, what functionality the scaffold serves cells.

In brief, future work is needed to invest in the sub-classifying philosophy and prove its efficacy. Despite the potential usefulness of the new categorizing method, it still needs to cooperate with existing classifying methods, at least in its current stage. In other word, the new and traditional methods are not mutually exclusive and could work together. Therefore, other proven scaffold-categorizing methods, such as categorizing via natural or synthetic materials, 3DP or traditional fabricating method, scaffold-based or scaffold-free, 3D solid construct or hydrogel, being templates or permanently useable, types of decellularized matrix from various tissues and organs, and so forth, still have significant value as far as scaffold categorization is concerned. These could possibly be restructured and integrated inside the evolution-based scaffold classifying system, hopefully making the general scaffold categorization work more comprehensive, effective, and specific.

4. Conclusions

This paper introduced the basic knowledge of cell culture scaffold, explaining the evolution line from static cell culture to dynamic cell culture, and from 2D cell culture to 3D. As scaffold engineering ascends into the cross-domain and miscellaneous, current mainstream classification of scaffold—for instance 'traditional' or 'novel'—needs to be more specific, scientific, and persuasive. A new categorization method for scaffold engineering is therefore necessary, in order to avoid misunderstanding between researchers, as well as to increase the working efficiency inside scaffold studies. In this connection, the novel categorization of scaffold from No. 1, 2, 3, and into 7 has been established, with a summarizing table concluded the state-of-the-art of seven-layer scaffold classification via the language of 'relativity'.

From another point of view, an exact understanding of the language is necessary. In this paper, efforts have been made to define scaffold based on 'relativity', which could make it possible at once to mention what scaffold is being discussed, from what point of view, and in what connection. The fundamental property of the new definition is that all concepts of 'scaffolds', as well as their mutual relationship, are concentrated around one idea. This idea is about a scaffold's 'evolution'. Therefore, seven-levels of scaffolds have been defined to potentially illustrate the whole evolution process of scaffold engineering that covers the past, present, and future. With the betterment in scaffold's categorization, future scaffold researchers might possibly be benefited, however unobvious, with increased clarity, precision, and understanding.

In terms of future works, researchers may choose to focus on several aspects. First is to further analyse the classification of the seven layers and its philosophy. Since the novel language is herein introduced from the philosophical realm into scaffold engineering, there may be some comprehension difficulties, however small, which can be further modified and improved. Secondly, investigating the sub-classifying inside each of the seven categorizations, which was pointed out at paper's end section, might also prove to be an interesting study. Solid sub-classifying could help the evolutional picture of scaffold engineering be more thorough. In addition, other works may involve introducing novel language of 'relativity' into other engineering realms beyond the scaffold, which might in turn contribute to novel classification art that is better and more scientific.

Funding: This research was supported in part by funds provided through the Australia Research Training Program (RTP).

Acknowledgments: I gratefully acknowledged Ke Xing as my supervisor and Hung-Yao Hsu as the co-supervisor, for their professional teaching and guide. I also want to thank Sunsayo Lu, George Gao, Peter Osky, Pasco Kim, Vista Sam, and Sanmori Kitch for their kind suggestions and emotional support during the research.

Conflicts of Interest: The author declares no conflict of interest.

References

1. Yuan, H.; Xing, K.; Hsu, H.-Y. Concept justification of future 3dpvs and novel approach towards its conceptual development. *Designs* **2018**, *2*, 23. [CrossRef]
2. Yuan, H.; Xing, K.; Hsu, H.-Y. Trinity of three-dimensional (3d) scaffold, vibration, and 3d printing on cell culture application: A systematic review and indicating future direction. *Bioengineering* **2018**, *5*, 57. [CrossRef] [PubMed]
3. Ouspenskii, P.D. *Tertium Organum*; Alfred A. Knopf: New York, NY, USA, 1981; Volumn 106, p. 798.
4. Santos, E.; Hernández, R.M.; Pedraz, J.L.; Orive, G. Novel advances in the design of three-dimensional bio-scaffolds to control cell fate: Translation from 2d to 3d. *Trends Biotechnol.* **2012**, *30*, 331–341. [CrossRef] [PubMed]
5. Greiner, A.M.; Richter, B.; Bastmeyer, M. Micro-engineered 3d scaffolds for cell culture studies. *Macromol. Biosci.* **2012**, *12*, 1301–1314. [CrossRef] [PubMed]
6. Bettinger, C.J. Biodegradable elastomers for tissue engineering and cell-biomaterial interactions. *Macromol. Biosci.* **2011**, *11*, 467–482. [CrossRef] [PubMed]
7. Choi, C.K.; Breckenridge, M.T.; Chen, C.S. Engineered materials and the cellular microenvironment: A strengthening interface between cell biology and bioengineering. *Trends Cell Biol.* **2010**, *20*, 705–714. [CrossRef] [PubMed]
8. Serra, T. 3d-printed biodegradable composite scaffolds for tissue engineering applications. Ph.D. Thesis, Universitat Politècnica de Catalunya, Barcelona, Spain, 2014.
9. Liu, F.; Liu, C.; Chen, Q.; Ao, Q.; Tian, X.; Fan, J.; Tong, H.; Wang, X. Progress in organ 3d bioprinting. *Int. J. Bioprint.* **2018**, *4*, 1–15. [CrossRef]
10. Ovsianikov, A.; Khademhosseini, A.; Mironov, V. The synergy of scaffold-based and scaffold-free tissue engineering strategies. *Trends Biotechnol.* **2018**, *36*, 348–357. [CrossRef] [PubMed]
11. Khoo, Z.X.; Teoh, J.E.M.; Liu, Y.; Chua, C.K.; Yang, S.; An, J.; Leong, K.F.; Yeong, W.Y. 3d printing of smart materials: A review on recent progresses in 4d printing. *Virtual Phys. Prototyp.* **2015**, *10*, 103–122. [CrossRef]
12. Bogue, R. Smart materials: A review of recent developments. *Assem. Autom.* **2012**, *32*, 3–7. [CrossRef]
13. Ning, L.; Chen, X. A brief review of extrusion-based tissue scaffold bio-printing. *Biotechnol. J.* **2017**, *12*, 1600671. [CrossRef] [PubMed]
14. Scarritt, M.E.; Pashos, N.C.; Bunnell, B.A. A review of cellularization strategies for tissue engineering of whole organs. *Front. Bioeng. Biotechnol.* **2015**, *3*, 43. [CrossRef] [PubMed]
15. Badylak, S.F.; Taylor, D.; Uygun, K. Whole-organ tissue engineering: Decellularization and recellularization of three-dimensional matrix scaffolds. *Annu. Rev. Biomed. Eng.* **2011**, *13*, 27–53. [CrossRef] [PubMed]
16. Crapo, P.M.; Gilbert, T.W.; Badylak, S.F. An overview of tissue and whole organ decellularization processes. *Biomaterials* **2011**, *32*, 3233–3243. [CrossRef] [PubMed]

17. Farran, A.J.; Teller, S.S.; Jia, F.; Clifton, R.J.; Duncan, R.L.; Jia, X. Design and characterization of a dynamic vibrational culture system. *J. Tissue Eng. Regen. Med.* **2013**, *7*, 213–225. [CrossRef] [PubMed]
18. Zhang, C.; Li, J.; Zhang, L.; Zhou, Y.; Hou, W.; Quan, H.; Li, X.; Chen, Y.; Yu, H. Effects of mechanical vibration on proliferation and osteogenic differentiation of human periodontal ligament stem cells. *Arch. Oral Biol.* **2012**, *57*, 1395–1407. [CrossRef] [PubMed]
19. Sun, J.; Tan, R. *Technology Assessment: Triz Technology System Evolution Theory*; World Scientific: Singapore, 2017; pp. 55–81.
20. Bukhman, I. *Triz Technology for Innovation*; Cubic Creativity Company: Tulsa, OK, USA, 2012.
21. Fourth dimension. *Encyclopedia of Occultism and Parapsychology*, 5th ed.; Melton, J.G., Ed.; Gale: Detroit, MI, USA, 2001; Volumn 1, p. 586.
22. Skobelev, V. On the possibility of detecting the fourth dimension of space in experiments in our three-dimensional subspace. *Russ. Phys. J.* **2015**, *57*, 1392–1397. [CrossRef]

© 2019 by the author. Licensee MDPI, Basel, Switzerland. This article is an open access article distributed under the terms and conditions of the Creative Commons Attribution (CC BY) license (http://creativecommons.org/licenses/by/4.0/).

Communication

Optimization of Decellularization Procedure in Rat Esophagus for Possible Development of a Tissue Engineered Construct

Panagiotis Mallis [1,†], Panagiota Chachlaki [1,†], Michalis Katsimpoulas [2], Catherine Stavropoulos-Giokas [1] and Efstathios Michalopoulos [1,*]

[1] Hellenic Cord Blood Bank, Biomedical Research Foundation Academy of Athens, 4 Soranou Ephessiou Street, 115 27 Athens, Greece
[2] Center of Experimental Surgery, Biomedical Research Foundation Academy of Athens, 4 Soranou Ephessiou Street, 115 27 Athens, Greece
* Correspondence: smichal@bioacademy.gr; Tel.: +30-2106597331; Fax: +30-210-6597345
† These authors were equally contributed to this work as first authors.

Received: 5 December 2018; Accepted: 20 December 2018; Published: 24 December 2018

Abstract: Background: Current esophageal treatment is associated with significant morbidity. The gold standard therapeutic strategies are stomach interposition or autografts derived from the jejunum and colon. However, severe adverse reactions, such as esophageal leakage, stenosis and infection, accompany the above treatments, which, most times, are life threating. The aim of this study was the optimization of a decellularization protocol in order to develop a proper esophageal tissue engineered construct. **Methods:** Rat esophagi were obtained from animals and were decellularized. The decellularization process involved the use of 3-[(3-cholamidopropyl) dimethylammonio]-1-propanesulfonate (CHAPS) and sodium dodecyl sulfate (SDS) buffers for 6 h each, followed by incubation in a serum medium. The whole process involved two decellularization cycles. Then, a histological analysis was performed. In addition, the amounts of collagen, sulphated glycosaminoglycans and DNA content were quantified. **Results:** The histological analysis revealed that only the first decellularization cycle was enough to produce a cellular and nuclei free esophageal scaffold with a proper extracellular matrix orientation. These results were further confirmed by biochemical quantification. **Conclusions:** Based on the above results, the current decellularization protocol can be applied successfully in order to produce an esophageal tissue engineered construct.

Keywords: esophagus; Barret's esophagus; decellularization; CHAPS; SDS; histological images; tissue engineered construct

---

1. Introduction

Esophageal disease-related morbidity has increased dramatically in the last 10 years. More than 10,000 people have been affected by various types of esophageal disorder, such as congenital or acquired esophageal diseases, esophageal atresia and esophageal trauma [1,2]. Furthermore, over 500,000 individuals are diagnosed with esophageal cancer each year, worldwide [3,4]. For early stage esophageal malignancies and Barret's esophagus, endoscopic mucosal resection (EMR) is the gold standard treatment [3]. However, most pathologies need segmental substitution of the esophagus with either autologous or synthetic grafts [5]. Autologous grafts from the stomach, jejunum or colon can be applied, but 40% of patients die due to serious adverse reactions, such as limited nutrition and esophagus infection [6].

Under this scope, a proper esophageal scaffold can be fabricated using the tissue engineering methods. Until now, many attempts have been performed in order to develop esophageal constructs utilizing polymer and synthetic materials, such as Dacron and expanded polytetrafluoroethylene (ePTFE) [5,7,8]. Unfortunately, these constructs are accompanied by severe complications, such as anastomic leakage and esophageal stenosis. The esophagus is an organ that is located behind the trachea and consists of the epithelium, mucosa, submucosa and muscularis propria. Reproduction of this complicated structure with polymers or synthetic materials or even with 3D printing is extremely difficult and this may be the primary reason for construct failure [5,7,8]. On the other hand, the use of an acellular esophageal scaffold may be more effective than the aforementioned attempts [5]. Indeed, decellularized matrices have been used successfully in the experimental and clinical setting in the past for tissue replacement of a wide variety of organs such as the trachea, bladder, arteries and veins. Decellularization aims to remove tissue resident cells, while preserving the extracellular matrix (ECM) of the organ, reducing, in this way, the immunogenicity of the produced material. In addition, decellularized matrices are characterized by having greater biocompatibility rather than the artificial scaffolds [9]. The future goal is the production of an *off the shelf* esophageal scaffold, which can properly be implanted to human patients.

Due to the complicated esophageal structure, the decellularization procedure must be established properly before the performance of any clinical attempts. Until now, a great effort has been performed by several groups worldwide in order to validate the decellularization procedure, but most of these studies have been accompanied by contradictory results regarding the preservation of ECM components in decellularized matrices [5,9–12]. In most of the above techniques, a combination of detergent and enzymatic treatments has been applied for successful cell removal. However, the esophagus is characterized by a collagen-rich ECM, which can be damaged by enzymes in an irreversible way. It is widely known that enzymes such as trypsin can cleave the collagen and elastin fibers, thus inducing severe damage to the tissue ECM. In addition, the extended use of anionic detergents such as sodium dodecyl sulfate (SDS) can damage the sulfated glycosaminoglycans (sGAGs), which are key components of the ECM. When crucial ECM components, such as collagen, elastin and sGAGs, are damaged, then the occurred decellularized constructs are characterized by totally different properties from the original ones; thus, their use as scaffold could be hampered [5,9–13].

The aim of this study was the validation of a previously described non-enzymatic decellularization protocol in rat esophagi (rES) [14,15]. For this purpose, decellularization of rES was evaluated after two cycles. Then, histological analysis, morphometric measurements and biochemical quantifications were performed. The future goal is the use of this protocol on esophagi derived from larger animals or cadaveric human donors to produce a proper esophageal tissue engineered construct that could be applied to the patients.

2. Materials and Methods

2.1. Preparation of Rat Esophagi

Esophagi were harvested under aseptic conditions from Sprague–Dawley rats (n = 30), weighing 250–300 g. All animals were provided by Biomedical Research Foundation Academy of Athens (BRFAA) and handled according to the guidelines of animal care which conform with the Helsinki declaration. In addition, this study was approved by the Bioethics Committee of BRFAA. Each harvested esophagus was rinsed in Phosphate Buffer Saline 1X (PBS 1X, Sigma-Aldrich, Darmstadt, Germany) and processed immediately to the decellularization procedure.

2.2. Decellularization of Rat Esophagi

rES ($n = 10$, $l = 4 \pm 1$ cm) were cut into 3 segments of 1 cm and submitted to decellularization buffers. The decellularization process was performed according to previous described protocols with some modifications [12,13]. Briefly the esophagus segments were subjected to CHAPS buffer at pH 7 (8 mM CHAPS, 1 M NaCl and 25 mM EDTA in PBS 1X, Sigma-Aldrich, Darmstadt, Germany) for 6 h at room temperature under constant agitation at 350 rpm. Then, the esophageal segments were subjected to SDS buffer at pH 7, (1.8 mM SDS, 1M NaCl and 25 mM EDTA in PBS 1X, Sigma-Aldrich, Darmstadt, Germany) for additional 6 h at room temperature under constant agitation at 350 rpm. Finally, the esophageal segments were incubated in α-Minimum Essentials Medium (α-MEM, Sigma-Aldrich, Darmstadt, Germany) supplemented with 40% v/v Fetal Bovine Serum (FBS, Sigma-Aldrich, Darmstadt, Germany) for 6 h at 37 °C under constant agitation at 350 rpm. The above procedure was repeated for 1 more cycle.

2.3. Histological Analysis

Native non decellularized ($n = 10$, $l = 1$ cm) and decellularized rES segments after the 1st ($n = 10$, $l \approx 1$ cm) and 2nd ($n = 10$, $l \approx 1$ cm) cycles were fixed in 4% v/v paraformaldehyde (PFA, Sigma-Aldrich, Darmstadt, Germany) for 4 h. Then, the samples were rehydrated, paraffin embedded and sectioned at 5 μm. The following histological stainings were performed in order to validate the effect of each decellularization cycle to the esophageal extracellular matrix. Hematoxylin and Eosin (H&E, Sigma-Aldrich, Darmstadt, Germany), Sirius Red (SR, Sigma-Aldrich, Darmstadt, Germany), Orcein Stain (OS, Sigma-Aldrich, Darmstadt, Germany) and Toluidine Blue (TB, Sigma-Aldrich, Darmstadt, Germany) were performed for the evaluation of the presence of cell nuclei, collagens, elastin and sGAGs, respectively. Images were acquired with a Leica DM L2 light microscope (Leica Microsystems, Weltzar Germany) and processed with Image J 1.46 (Wane Rasband, National Institute of Health, Bethesda, MD, USA).

Indirect immunofluorescence against fibronectin in combination with DAPI was performed in native and decellularized esophageal segments. Briefly, the slides were deparaffinized, rehydrated and blocked. Then, monoclonal antibody against rat fibronectin (1:5000, Sigma-Aldrich, Darmstadt, Germany) was added, incubated and followed by the addition of secondary FITC- conjugated mouse IgG antibody (1:100, Sigma-Aldrich, Darmstadt, Germany). Finally, DAPI was added in slides and incubated. The slides were mounted with glycerol and observed under a Leica SP5 II fluorescence microscope equipped with LAS Suite v2 software (Leica Microsystems, Weltzar, Germany).

2.4. Morphometric Analysis

The morphometric analysis involved the measurement of length, mucosa thickness and total thickness in native and decellularized rES. Specifically, the length and total thickness were measured in native ($n = 10$) and decellularized rES after the 1st ($n = 10$) and 2nd ($n = 10$) cycles. Mucosa thickness was determined only in native ($n = 10$) and decellularized ($n = 10$) rES after 1st cycle. After the 2nd decellularization cycle, the rES ECM was damaged and thus, it could not be efficiently used to estimate mucosa thickness. The length of rES was determined with a digital caliper (Flip-Plus Electronic Caliper, Fowler, Newton, MA, USA). Mucosa thickness and total thickness measurements were performed in histological images, using Image J 1.46 (Wane Rasband, National Institute of Health, Bethesda, MD, USA).

2.5. Quantification of Collagen, sGAGs and DNA Content

Native ($n = 10$, $l = 1$ cm) and decellularized esophagi segments after the 1st ($n = 10$, $l \approx 1$ cm) and 2nd ($n = 10$, $l \approx 1$ cm) cycles were digested using a lysis buffer contained 0.1 M Tris pH 8, 0.2 M NaCl and 5 mM EDTA in PBS 1X (Sigma-Aldrich, Darmstadt, Germany) supplemented with 30 mg/mL Proteinase K (Sigma-Aldrich, Darmstadt, Germany). The digestion was performed at 56 °C for 12 h followed by inactivation at 90 °C for 5 min.

The amount of collagen in each sample was quantified with a Hydroxyproline Assay kit (MAK008, Sigma-Aldrich, Darmstadt, Germany) according to the manufacturer's instructions. SGAGs were measured by the addition of 1% w/v dimethylene blue (Sigma-Aldrich, Darmstadt, Germany), and photometric measurement was done at 525 nm. The concentration of sGAGs in each sample was obtained through interpolation to a standard curve. A standard curve was developed based on the chondroitin sulfate standards of 12 μg/mL, 25 μg/mL, 50 μg/mL, 100 μg/mL and 150 μg/mL.

Finally, for the DNA quantification assay, the total genetic material from each sample was eluted in 100 μL of DNAse free water (Sigma-Aldrich, Darmstadt, Germany) followed by spectrophotometric quantification at 260 to 280 nm.

2.6. Statistical Analysis

Graph Pad Prism v 6.01 (GraphPaD Software, San Diego, CA, USA) was used for the statistical analysis. All parameters of this study were compared, using the Mann–Whitney test. Statistically significant differences between group values were considered when the p-value was less than 0.05. Indicated values are presented as mean ± standard deviation.

3. Results

3.1. Histological Analysis

rES were successfully decellularized with the current protocol. After the first decellularization cycle, rES were characterized by well-preserved ECM, while cellular and nuclear materials were eliminated (Figure 1). Decellularized rES after the second cycle presented extensive damage in their ECMs.

Specifically, decellularized rES after the first cycle successfully retained their matrix components, such as collagen, elastin and sGAGs, as indicated by SR, OS and TB stains (Figure 1). Indeed, the SR stain revealed the preservation of the collagen fibers in decellularized rES after the first cycle. Moreover, decellularized rES appeared to have a more compact structure when compared to native samples. This phenomenon might be a result of cell loss during the decellularization procedure. Elastin fibers were stained black by OS, thus revealing their intact structure in decellularized rES after the first cycle. Finally, decellularized rES after the first cycle were characterized by a weaker TB stain as compared with the native rES.

On the other hand, decellularized rES after the second cycle, presented wide destruction of ECM key components. Indeed, the inner layer of decellularized rES after the second cycle was totally detached from the rest of the esophageal matrix (Figure 1). In addition, collagen fibers appeared to be damaged in decellularized rES after the second cycle, as was indicated by the weaker SR stain when compared to native samples. Elastin was absent in decellularized esophagi after the second cycle, whereas sGAGs did not present any significant alteration.

Indirect immunofluorescence results showed the preservation of well oriented fibronectin in rES after the first cycle (Figure 1 and Figure S1). RES from the second decellularization cycle were characterized by damaged fibronectin (Figure 1). DAPI stain was observed only in native samples. No DAPI stain was evident in decellularized rES either from the first or second cycle (Figure 1). The indirect immunofluorescence results regarding the preservation or damage of ECM components appeared to be consistent with the histological staining results. The above results strongly indicate that one cycle is enough to produce a proper decellularized rES without further damage to its ultrastructure.

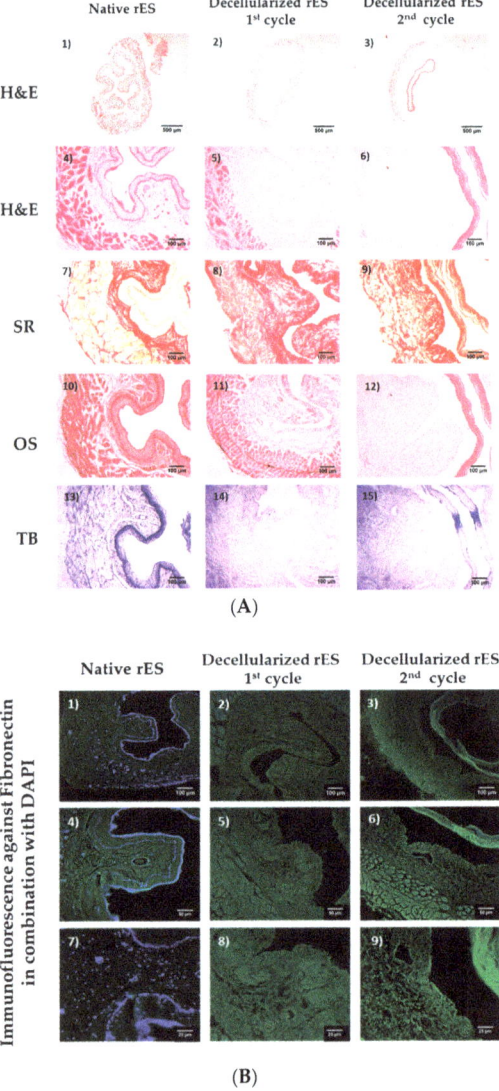

Figure 1. Histological analysis of decellularized rat esophagi (rES) after the first and second cycles. Native rES stained with H&E (**A1,A4**), SR (**A7**), OS (**A10**) and TB (**A13**). Decellularized rES stained with H&E (**A2,A3,A5,A6**), SR (**A8,A9**), OS (**A11,A12**) and TB (**A14,A15**) after the first and second cycles, respectively. The black arrows indicate elastin preservation in decellularized esophagi after the first cycle. Images A1–A3 are presented with original magnification 2.5×; scale bars are 500 µm. Images A4–A15 are presented with original magnification 10×; scale bars are 100 µm. Indirect immunofluorescence against fibronectin in combination with DAPI in native (**B1,B4,B7**) and decellularized rES after the first (**B2,B5,B8**) and second (**B3,B6,B9**) cycles. Images B1–B3 are presented with original magnification 10×; scale bars are 100 µm. Images B4, B5 and B6 are presented with original magnification 20×; scale bars are 50 µm. Images B7, B8 and B9 are presented with original magnification 40×; scale bars are 25 µm.

3.2. Morphometric Analysis

Further validation of the current decellularization protocol to rES involved a morphometric analysis. For this purpose, histological images were used in order to measure the mucosa thickness and the total thickness, while the total esophageal length was measured with a digital caliper.

The length of decellularized rES after the first cycle reduced by 19% and after the second cycle by 36%. Specifically, the length of native non-decellularized rES was 1.0 ± 0.1 cm, while the length of decellularized rES after the first and second cycles was 0.8 ± 0.1 cm and 0.6 ± 0.1 cm, respectively (Figure 2). This decrease in length between native and decellularized esophagi from both cycles was found to be statistically significant ($p < 0.001$).

Native rES consists of epithelium, mucosa, submucosa and muscularis propria (Figure S2). Among them, mucosa thickness and total thickness were measured.

Specifically, the total thickness was 0.5 ± 0.1 mm in native rES, 0.4 ± 0.1 mm in decellularized rES after the first cycle and 0.3 ± 0.1 cm after the second cycle (Figure 2). Statistically significant differences were observed in total thickness between the native and decellularized rES after the first ($p < 0.05$) and second cycles ($p < 0.001$). Furthermore, statistically significant differences were observed in the total thickness between decellularized rES of the first and second cycles ($p < 0.05$). Finally, the thickness of mucosa layer in native and decellularized rES was measured. The thickness of mucosa layer was 65 ± 1 μm in native rES and 15 ± 1 μm in decellularized rES after the first cycle, while this layer totally damaged after the second decellularization cycle and could not be measured (Figure 2).

3.3. Biochemical Analysis and DNA Quantification

Collagen, sGAG and DNA content were quantified in order to validate the current decellularization protocol. Specifically, native rES characterized by 55.1 ± 7.3 μg hydroxyproline per mg of dry tissue weight, while decellularized rES after the first and second cycles were characterized by 53.2 ± 5.5 and 28.1 ± 6.4 μg hydroxyproline per mg of dry tissue weight, respectively (Figure 2). No statistically significant difference was observed in the collagen amount between native and decellularized rES after the first cycle. Statistically significant differences were observed between native and decellularized rES after the second cycle ($p < 0.001$) and decellularized rES from the first to the second cycle ($p < 0.05$).

SGAG content was significantly reduced between native and decellularized rES from both cycles. The SGAG content in native rES was 2.4 ± 0.5 μg sGAG per mg tissue weight. Decellularized rES after the first and second cycles were characterized by 0.5 ± 0.2 and 0.4 ± 2 μg sGAG per mg of dry tissue weight, respectively (Figure 2).

The DNA amount in native rES was 1300 ± 268 ng DNA per μg of dry tissue weight, while after the first decellularization cycle, it was 93 ± 26 ng DNA per μg of dry tissue weight and after the second decellularization cycle, it was 93 ± 29 ng DNA per μg of dry tissue weight (Figure 2). Statistically significant differences were observed between native and decellularized rES either by the first ($p < 0.001$) or second cycle ($p < 0.001$).

4. Discussion

Esophagus-related diseases and their postoperatively complications are associated with a high mortality rate [1–5]. These diseases affect either pediatric or adult patients and, most times, the use of autografts derived either from jejunum or gastric interposition or synthetic grafts are the gold standard treatments [3–5]. Unfortunately, severe complications are frequently observed and can be life threating for the patients.

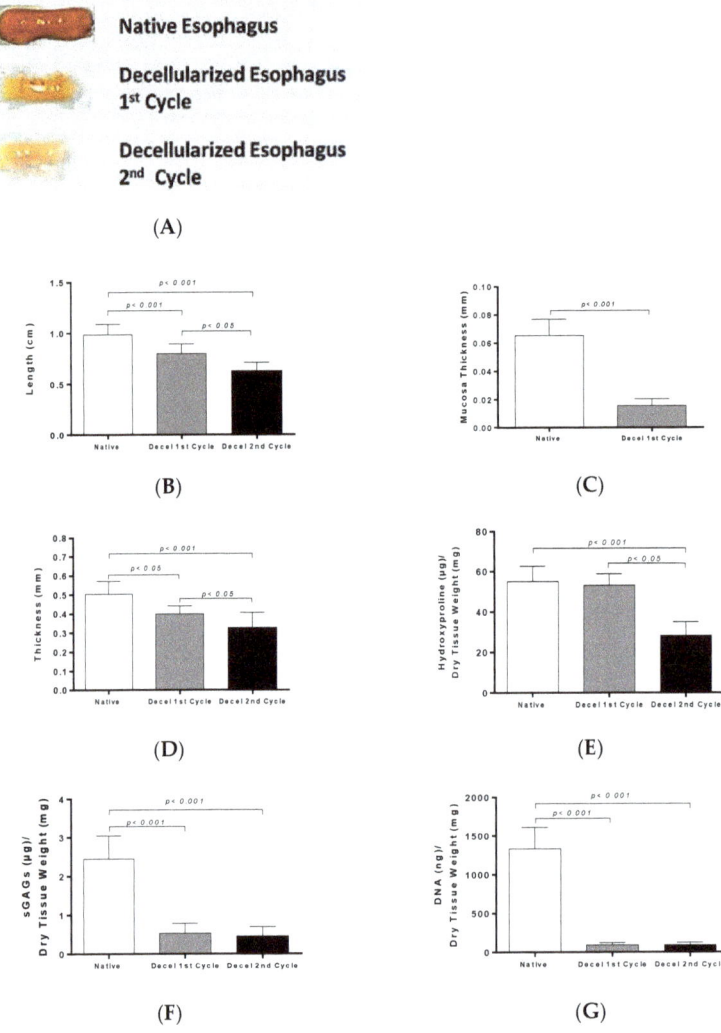

Figure 2. Morphometric analysis and biochemical quantification in rES. Macroscopic overview of native and decellularized rES after the first and second cycles (**A**). Measurement of length (**B**), mucosa thickness (**C**) and total thickness (**D**) in rES. Statistically significant differences were observed in the above parameters between native and decellularized rES either by the first ($p < 0.005$) or second cycle ($p < 0.001$); Biochemical quantification involved hydroxyproline measurement (**E**); sulfated glycosaminoglycans (sGAGs) (**F**) and DNA (**G**) content determination. Statistically significant differences were observed in collagen, sGAG and DNA content between native and decellularized rES either by the first ($p < 0.05$) or second cycle ($p < 0.05$).

Esophagus tissue engineering is a promising solution, although it is still challenging, and more effort must be performed in this direction. The aim of this study was to validate a decellularization protocol that has been previously used successfully in other tissues, such as human umbilical arteries and porcine pericardium [14–16]. The goal of this study was to produce a proper esophageal scaffold, reducing the ECM damage and the processing time. For this purpose, histological analysis, morphometric measurements and quantification of collagen, sGAGs and DNA were performed.

The histological analysis revealed the successful decellularization of rES using only one decellularization cycle. Indeed, after the first decellularization cycle, rES were characterized by a properly organized ECM, collagen, elastin and sGAGs were retained, while total cellular and nuclear materials were lacking. However, after the second decellularization cycle, rES were extensively damaged. The inner mucosa layer appeared to be detached from the rest of the esophageal matrix. Moreover, these results are similar to previous reports from other groups, thus indicating the success of the current decellularization protocol [9–11,17,18]. In most of these studies, an increased number of decellularization cycles or increased processing time was required. In our study, only one decellularization cycle was needed to successfully decellurize the rES. Specifically, in the study of Urbani et al. [18], three decellularization cycles were used in order to completely remove the cellular and nuclear materials. In Urbani's study, even after the first decellularization cycle, the esophagus was characterized by damaged collagen and elastin fibers. Unlike these results, in our experimental procedure, the histological analysis indicated well-organized esophageal ECMs with no cellular and nuclear materials and without the need for a second decellularization cycle. This discrepancy in the results between these two studies might be attributed to the different origins of the esophagi and decellularization protocol that were used.

The histological analysis also involved indirect immunofluorescence against fibronectin in combination with DAPI staining in native and decellularized rES. Fibronectin appeared to be well preserved in rES after the first decellularization cycle, while the rES of the second cycle were characterized by damaged fibronectin. No DAPI stain was evident in decellularized rES, thus further confirming our initial histological results regarding the absence of nuclear materials. These results are in accordance with the study by Bhrany et al., indicating the successful preservation of fibronectin after decellularization [9]. Fibronectin plays a significant role in the epithelium maintenance and function through its binding sites. The preservation of fibronectin in decellularized rES is of major importance, as it makes them efficient for re-epithelization.

The morphometric analysis revealed that the length, mucosa thickness and total thickness of the rES were significantly decreased after each decellularization cycle. As a consequence of these morphometric changes, biomechanical alterations may be revealed, as has been reported in previous studies [9,11]. In a number of studies including decellularized matrices from various origins, such as vessels and aortic valves, the thickness was increased [10,17,18]. In most of these studies, thickness measurements were performed with caliper instruments in non-formalin fixed decellularized matrices. It is known that decellularized matrices attract water molecules, thus increasing their total weights and enlarging their wall thicknesses. In order to validate the thickness change between native and decellularized tissues, the initial water content of native tissues must be determined. Any attempt to perform the above measurements in non-fixed native and decellularized matrices will be not be accurate enough. In order to avoid this phenomenon, in our study, mucosa and total thickness were measured from histological images using image analysis software. In this way, the same dehydration rate was achieved both in native and decellularized tissues, obtaining, in this way, more accurate results than the aforementioned studies [10,18,19].

The next step of this study was biochemical quantification which involved collagen, sGAGs and DNA determination. The collagen content did not present any statistically significant changes between native and decellularized rES after the first and second cycles. On the contrary, sGAGs were decreased after the first decellularization cycle, followed by an extensive decrease after the second decellularization cycle. This decrease in sGAG content between native and decellularized rES (first and second cycles) was statistically significant. SGAGs form large macromolecules called proteoglycans and are responsible for the collagen orientation in tissues. SDS, a reagent that was used in the decellularization protocol, can harm the sGAGs through binding to their negatively-charged sites. This decrease in sGAG content might alter the collagen orientation, thus damaging the tissue ECM. However, no structural alterations of ECM in decellularized rES after the first cycle were observed as indicated by the histological analysis. The wide damage that was observed in the ECM of rES after the second cycle, was possibly due to the greater impact of the decellularization reagents to all structural tissue components (collagen, elastin, fibronectin and sGAGs) and not specifically to sGAGs.

Further confirmation of the successful removal of genetic material from decellularized rES was performed by DNA quantification. DNA content was reduced by over 95% in decellularized rES from both cycles. The above results are comparable with previous works in other tissues and further confirm that one decellularization cycle is more than enough to produce an esophageal tissue engineered construct.

Under this scope, decellularized rES was characterized by a properly preserved ECM with no cellular or nuclear material, as indicated by the histological stains. Moreover, histological and biochemical analysis results were found to be consistent with the criteria of successful tissue decellularization that were provided by the study of Crapo et al. [13,20]. In this way, and based on the above results, strong evidence is provided regarding the successful decellularization of rat esophagi with the proposed decellularization protocol [13,20].

5. Conclusions

The aim of this study was to optimize a decellularization protocol for proper development of an esophageal tissue engineered construct. Unlike previous published studies [9–11,15,16], our decellularization protocol is cost effective and less time consuming, producing a decellularized matrix with the same structure and function properties. In order to further confirm the proper preservation of ECM components and their properties as an esophageal scaffold, more experiments need to be performed, including a cytotoxicity assay, biomechanical testing, recellularization with tissue specific cell populations (epithelial cells and muscle cells) and implantation to animal models. The future goal of this study is the use the of current decellularization protocol in esophagi from larger animal models and from cadaveric human donors in order to develop a proper esophageal tissue engineered construct. This construct could be applied in patients, eliminating the use of autologous stomach, intestine conduits or synthetic grafts, thus lowering the morbidity which is caused by the adverse side effects of the above applications.

Supplementary Materials: The following are available online at http://www.mdpi.com/2306-5354/6/1/3/s1, Figure S1: Indirect immunofluorescence against fibronectin in combination with DAPI stain in native and decellularized rES. Figure S2: Histological image of native and decellularized rat esophagus with H&E.

Author Contributions: P.M. (first author) and P.C. (equal first author) carried out the whole experimental procedure of this study. In addition, P.M. performed the statistical analysis. M.K. was the person responsible for the animal care and esophagus harvesting. C.S.-G. supervised and approved the overall study. E.M. supervised and approved the study.

Funding: This research received no funding.

Conflicts of Interest: The authors declare no conflict of interest.

References

1. Bray, F.; Ferlay, J.; Soerjomataram, I.; Siegel, R.L.; Torre, L.A.; Jemal, A. Global cancer statistics 2018: GLOBOCAN estimates of incidence and mortality worldwide for 36 cancers in 185 countries. *CA Cancer J. Clin.* **2018**, *68*, 394–424. [CrossRef]
2. Lambert, R.; Hainaut, P. The multidisciplinary management of gastrointestinal cancer. Epidemiology of oesophagogastric cancer. *Best Pract. Res. Clin. Gastroenterol.* **2007**, *21*, 921–945. [CrossRef]
3. Napier, K.J.; Scheerer, M.; Misra, S. Esophageal cancer: A Review of epidemiology, pathogenesis, staging workup and treatment modalities. *World J. Gastrointest. Oncol.* **2014**, *15*, 112–120. [CrossRef]
4. Clark, D.C. Esophageal atresia and tracheoesophageal fistula. *Am. Fam. Physician* **1999**, *59*, 910–916. [PubMed]
5. Totonelli, G.; Maghsoudlou, P.; Fishman, J.M.; Orlando, G.; Ansari, T.; Sibbons, P.; Birchall, M.A.; Pierro, A.; Eaton, S.; De Coppi, P. Esophageal tissue engineering: A new approach for esophageal replacement. *World J. Gastroenterol.* **2012**, *21*, 6900–6907. [CrossRef] [PubMed]

6. Egorov, V.I.; Schastlivtsev, I.V.; Prut, E.V.; Baranov, A.O.; Turusov, R.A. Mechanical properties of the human gastrointestinal tract. *J. Biomech.* **2002**, *35*, 1417–1425. [CrossRef]
7. Freud, E.; Efrati, I.; Kidron, D.; Finally, R.; Mares, A.J. Comparative experimental study of esophageal wall regeneration after prosthetic replacement. *J. Biomed. Mater. Res.* **1999**, *45*, 84–91. [CrossRef]
8. Lynen Jansen, P.; Klinge, U.; Anurov, M.; Titkova, S.; Mertens, P.R.; Jansen, M. Surgical mesh as a scaffold for tissue regeneration in the esophagus. *Eur. Surg. Res.* **2004**, *36*, 104–111. [CrossRef] [PubMed]
9. Bhrany, A.D.; Beckstead, B.L.; Lang, T.C.; Farwell, D.G.; Giachelli, C.M.; Ratner, B.D. Development of an esophagus acellular matrix tissue scaffold. *Tissue Eng.* **2006**, *12*, 319–330. [CrossRef] [PubMed]
10. Saxena, A.K.; Baumgart, H.; Komann, C.; Ainoedhofer, H.; Soltysiak, P.; Kofler, K.; Höllwarth, M.E. Esophagus tissue engineering: In situ generation of rudimentary tubular vascularized esophageal conduit using the ovine model. *J. Pediatr. Surg.* **2010**, *45*, 859–864. [CrossRef] [PubMed]
11. Luc, G.; Charles, G.; Gronnier, C.; Cabau, M.; Kalisky, C.; Meulle, M.; Bareille, R.; Roques, S.; Couraud, L.; Rannou, J.; et al. Decellularized and matured esophageal scaffold for circumferential esophagus replacement: Proof of concept in a pig model. *Biomaterials* **2018**, *175*, 1–18. [CrossRef] [PubMed]
12. Urbani, L.; Camilli, C.; Phylactopoulos, D.E.; Crowley, C.; Natarajan, D.; Scottoni, F.; Maghsoudlou, P.; McCann, C.J.; Pellegata, A.F.; Urciuolo, A.; et al. Multi-stage bioengineering of a layered oesophagus with in vitro expanded muscle and epithelial adult progenitors. *Nat. Commun.* **2018**, *16*, 4286. [CrossRef] [PubMed]
13. Gilbert, T.W.; Sellaro, T.L.; Badylak, S.F. Decellularization of tissues and organs. *Biomaterials* **2006**, *27*, 3675–3683. [CrossRef] [PubMed]
14. Gui, L.; Chan, S.A.; Breuer, C.K.; Niklason, L.E. Novel utilization of serum in tissue decellularization. *Tissue Eng. Part C Methods* **2010**, *16*, 173–184. [CrossRef] [PubMed]
15. Mallis, P.; Gontika, I. Evaluation of decellularization in umbilical cord artery. *Transplant. Proc.* **2014**, *46*, 3232–3239. [CrossRef] [PubMed]
16. Mallis, P.; Michalopoulos, E.; Dimitriou, C.; Kostomitsopoulos, N.; Stavropoulos-Giokas, C. Histological and biomechanical characterization of decellularized porcine pericardium as a potential scaffold for tissue engineering applications. *Biomed. Mater. Eng.* **2017**, *28*, 477–488. [CrossRef] [PubMed]
17. La Francesca, S.; Aho, J.M.; Barron, M.R.; Blanco, E.W.; Soliman, S.; Kalenjian, L.; Hanson, A.D.; Todorova, E.; Marsh, M.; Burnette, K.; et al. Long-term regeneration and remodeling of the pig esophagus after circumferential resection using a retrievable synthetic scaffold carrying autologous cells. *Sci. Rep.* **2018**, *7*, 4123. [CrossRef] [PubMed]
18. Urbani, L.; Maghsoudlou, P.; Milan, A.; Menikou, M.; Hagen, C.K.; Totonelli, G.; Camilli, C.; Eaton, S.; Burns, A.; Olivo, A.; et al. Long-term cryopreservation of decellularised oesophagi for tissue engineering clinical application. *PLoS ONE* **2017**, *9*, e0179341. [CrossRef] [PubMed]
19. Granados, M.; Morticelli, L.; Andriopoulou, S.; Kalozoumis, P.; Pflaum, M.; Iablonskii, P.; Glasmacher, B.; Harder, M.; Hegermann, J.; Wrede, C.; et al. Development and Characterization of a Porcine Mitral Valve Scaffold for Tissue Engineering. *J. Cardiovasc. Transl. Res.* **2017**, *10*, 374–390. [CrossRef] [PubMed]
20. Crapo, P.M.; Gilbert, T.W.; Badylak, S.F. An overview of tissue and whole organ decellularization processes. *Biomaterials* **2011**, *32*, 3233–3243. [CrossRef] [PubMed]

© 2018 by the authors. Licensee MDPI, Basel, Switzerland. This article is an open access article distributed under the terms and conditions of the Creative Commons Attribution (CC BY) license (http://creativecommons.org/licenses/by/4.0/).

Article

Decellularized Human Umbilical Artery Used as Nerve Conduit

Ioanna Gontika [1], Michalis Katsimpoulas [2], Efstathios Antoniou [3], Alkiviadis Kostakis [2], Catherine Stavropoulos-Giokas [2] and Efstathios Michalopoulos [1,*]

[1] Hellenic Cord Blood Bank, Biomedical Research Foundation Academy of Athens, 4 Soranou Ephessiou Street, 11527 Athens, Greece
[2] Center of Clinical, Experimental Surgery and Translational Research, Biomedical Research Foundation of the Academy of Athens, 4 Soranou Ephessiou Street, 11527 Athens, Greece
[3] Second Department of Propaedeutic Surgery, University of Athens, Medical School, "Laiko" General Hospital 17 Agios Thomas Street, 11527 Athens, Greece
* Correspondence: smichal@bioacademy.gr; Tel.: +30-210-6597-331; Fax: +30-210-6597-345

Received: 28 September 2018; Accepted: 16 November 2018; Published: 21 November 2018

Abstract: Treatment of injuries to peripheral nerves after a segmental defect is one of the most challenging surgical problems. Despite advancements in microsurgical techniques, complete recovery of nerve function after repair has not been achieved. The purpose of this study was to evaluate the use of the decellularized human umbilical artery (hUA) as nerve guidance conduit. A segmental peripheral nerve injury was created in 24 Sprague–Dawley rats. The animals were organized into two experimental groups with different forms of repair: decellularized hUA ($n = 12$), and autologous nerve graft ($n = 12$). Sciatic faction index and gastrocnemius muscle values were calculated for functional recovery evaluation. Nerve morphometry was used to analyze nerve regeneration. Results showed that decellularized hUAs after implantation were rich in nerve fibers and characterized by improved Sciatic Functional index (SFI) values. Decellularized hUA may support elongation and bridging of the 10 mm nerve gap.

Keywords: umbilical arteries; nerve regeneration; nerve conduit

1. Introduction

Peripheral Nervous injuries (PNI) are a global clinical problem, since they significantly affect the quality of life of patients and cause enormous socio-economic burden [1–3]. Indicatively, in the United States alone, more than 50,000 peripheral nerve repair surgeries are performed annually [4]. The use of autologous nerve graft is considered as the gold standard procedure for bridging peripheral nerve defects. However, surgical approaches are characterized by several drawbacks. For instance, a secondary surgical procedure, which is often associated with donor site pain and mobility, is needed in order to obtain the nerve graft. Moreover, the sources of neural tissue that can be used as nerve conduits, are particularly limited [5–8]. Other approaches include the development of three-dimensional scaffold-nerve conduits which can be used for gap bringing between the proximal and distal stump of the nerve tissue. In many cases, nerve conduits also act as cells or growth factors carriers [9,10]. Nerve conduits have been fabricated using different types of material; natural and synthetic, biodegradable and non-biodegradable [11,12]. Natural biological nerve conduits such as vessels (veins and arteries), decellularized nerve [13] and muscle tissue have been widely used to bridge peripheral nerve gap in animal models [14] and also in clinical practice [15,16]. Tissue decellularization offers the possibility to obtain a cell-free, natural extracellular matrix (ECM), characterized by an adequate 3D organization with proper composition to repair different tissues or organs, including peripheral nerves [13].

The human umbilical cord contains two arteries, which can easily be isolated without invasive procedures. Previous reports showed that hUAs retained after decellurization their components such as collagen type I, laminin, and fibronectin [17]. These ECM components are also represented in the ECM of the peripheral nerve [18–20]. The aim of this study was to evaluate the use of the decellularized hUA as a nerve guidance conduit in a rat sciatic nerve model.

2. Materials and Methods

2.1. Collection and Isolation of Human Umbilical Arteries

Human umbilical cords were collected after informed consent from healthy donors. The informed consent was in accordance with Helsinki declaration and approved by the ethical committee of Biomedical Research Foundation Academy of Athens (BRFAA). The cords were stored at 4 °C immediately after birth and the overall storage time until processing did not exceed 24 h. Arteries were isolated from the cords using sterile surgical tools followed by brief rinses in Phosphate Buffer Saline 1× (PBS 1×).

2.2. Decellularization of Human Umbilical Arteries

The decellularization of the hUAs was carried out according to previously described protocols, [17]. Briefly, hUAs ($n = 44$, $l = 2$ cm), were incubated in CHAPS solution (8 mM CHAPS (APPLICHEM, Darmstadt, Germany), 1 M NaCl, and 25 mM EDTA (Ethylenediaminetetraacetic acid) in PBS 1×; (Sigma-Aldrich, Darmstadt, Germany) at pH 8 for 22 h, followed by brief washes in PBS 1×. The hUA were further incubated in SDS solution (1.8 mM SDS (Sigma-Aldrich, Darmstadt, Germany), 1 M NaCl, and 25 mM EDTA in PBS 1×) at pH 7.5 for 24 h, followed by 3 washes for 5 min in PBS 1×, to completely remove the detergent. Finally, the arteries were incubated at 37 °C for 48 h in alpha- Minimal Essential Medium (α-MEM, Gibco Life Technology, Darmstadt Germany), containing 40% (v/v) fetal bovine serum (FBS, Gibco Life Technology, Darmstadt, Germany) and 1000 U/mL penicillin-streptomycin (Gibco Life Technology, Darmstadt, Germany). All steps were performed under agitation and sterile conditions.

2.3. Histological Evaluation of Decellularized Arteries

Native and decellularized hUAs ($n = 10$) were fixed overnight in 10% v/v neutral buffered formalin, embedded in paraffin, cut into 5 μm sections and finally stained with Hematoxylin and Eosin (H&E, Sigma-Aldrich, Darmstadt, Germany) for nuclear material, Masson's Trichrome (Sigma-Aldrich, Darmstadt, Germany) for collagen content.

2.4. Evaluation of Toxicity of Decellularized Umbilical Arteries

2.4.1. Contact Cytotoxicity Assay

The decellularized hUAs ($n = 10$) were cut into 5 × 5 mm and placed in a 24 well culture plate (Orange Scientific, Braine-l'Alleud, Belgium). MSCs (Mesenchymal stem cell) were isolated from Wharton's Jelly tissue and were seeded into each well at a density of 1×10^4 cells. Then, the samples were incubated at 37 °C in 5% (v/v) CO_2 for 48 h. As positive control group for this assay, SDS was added in MSCs ($n = 10$), and as negative control group MSCs ($n = 10$) were cultured under normal conditions. Morphological examination of seeded cells was performed using brightfield microscope (LEICA DM 1L, Wetzlar, Germany). Images were captured using IC Capture 2.2 software.

2.4.2. ADP/ATP Ratio Assay

Native ($n = 20$) and decellularized hUAs ($n = 20$) were digested using lysis buffer, consisted of 1 mL α-MEM with 1 mg/mL Proteinase K (Sigma-Aldrich, Darmstadt, Germany). The digestion was performed overnight at 56 °C, and the following day the Proteinase K was inactivated at 95 °C for 5 min. The lysates from native and decellularized hUAs were used as culture medium for the evaluation of metabolic activity in MSCs. Then, 1×10^3 MSCs were adhered to each well of 96-wells plate, and the above lysates were added. Specifically, lysate derived from native hUAs was added in 10 wells with adhered MSCs. Lysates derived from decellularized hUAs were added to the next 10 wells with adhered MSCs. MSCs cultured with 1.2 mM SDS (Sigma-Aldrich, Darmstadt, Germany) were used as positive control group. As negative control was used MSCs cultured with standard medium. The culture medium was consisted of α-MEM ((Sigma-Aldrich, Darmstadt, Germany) supplemented with 15% v/v FBS (Sigma-Aldrich, Darmstadt, Germany) and 1% v/v Penicillin (Sigma-Aldrich, Darmstadt, Germany) and 1% v/v Streptomycin (Sigma-Aldrich, Darmstadt, Germany). The 96-well plate was incubated at 37 °C in 5% (v/v) CO_2 for 24 h. Subsequently ADP/ATP ratio assay (Sigma-Aldrich Ratio Assay Kit) was performed according to manufacturer's instructions.

2.5. Animals

Twenty-four male Sprague–Dawley (DS) rats, weighting 250–300 g were randomly divided into two groups (n = 12 in each group): The first group was consisted of decellularized hUAs and compared with the second group, which consisted of nerve autograft. The animals were provided by the Animal center of BRFAA and were handled in compliance with the guidelines for the use and care of laboratory animals. Furthermore, all animals were kept in a temperature-controlled room with a 12/12-h light/dark cycle and provided with rodent diet and water ad libitum. The study protocol was approved by the general veterinary directorate and animal health directorate with reference number 2777/26-04-2016 and was accepted by the Bioethics Committee of BRFAA

2.6. Surgical Procedure

The animals were anesthetized by isoflurane 3% in 1 L of oxygen. A dorsal gluteal-splitting approach was used to expose and mobilize the right sciatic nerve of each animal. The right sciatic nerve was exposed and a 1 cm gap was made in the mid portion of the nerve. In the nerve autograft group, the removed segment of nerve was oriented at 180° and grafted into the same nerve gap with 6 stiches of prolene 8-0 sutures. In the umbilical artery group, a 1.5 cm artery was grafted into the gap. Both proximal and distal stump were inserted about 2–3 mm from the ends of the artery graft and four stiches were performed in each stump. The manipulations of the nerves were made under an operational microscope.

2.7. Sciatic Functional Index (SFI)

The functional condition of the animals was assessed with the estimation of SFI, according to Bain et al. Formula [21]:

$$-38.3 \frac{EPL - NPL}{NPL} + 109.5 \frac{ETS - NTS}{NTS} + 13.3 \frac{EIT - NIT}{NIT} - 8.8.$$

Walk track analysis was performed at pre-operative, and at the 4th and 12th week after the surgery [22]. The rats' hind feet were painted with ink and the animals were placed in a walking pathway to walk down the track, leaving their footprint. Footprints from the experimental (E) and contralateral normal (N) sides were analyzed by measuring the lengths of the third toe to heel (PL), the first toe to the fifth toe (TS), and the second toe to the fourth toe (IT). Index values close to 0 indicated normal function and values close to -100 represented loss of function.

2.8. Nerve Graft Harvested Tissue

Twelve weeks postoperatively, the regenerated sciatic nerves were harvested. The midportion of the graft n = 6 from each group were fixed with 10% neutral buffered formalin solution, for immunohistochemical analysis. In addition, the other n = 6 from each group were fixed with 2.5% glutaraldehyde in 0.1 M phosphate buffer for morphometric analysis. Transverse sections were cut both in immunochemistry and in morphometry. The sections which were analyzed in this assay, were 5 mm distal from the side of proximal lesion.

2.9. Nerve Immunohistochemistry

Grafts were fixed in 10% v/v neutral buffer formalin solution (Sigma-Aldrich, Darmstadt, Germany) paraffin embedded and sectioned. Then, the slides were deparafinnized, rehydrated and blocked. Dako Envision Flex kit was used for the immunohistochemistry assay according to manufacturer's instructions (Dako, Agilent, Glostrup, Denmark). Briefly, nerve graft sections were incubated at 4 °C over night with rabbit anti-neurofilament 200 (nf 200) antibody (1:80, Sigma, St. Louis, MO, USA) to identify axons and S100 antibody (1:100, Sigma, St. Louis, MO, USA) to identify Schwann cells. Briefly, washes were performed, and addition of horseradish peroxidase (HRP) conjugated with goat secondary antibody against rabbit and mouse was performed. The slides were incubated at Room Temperature (RT) for 45 min. Finally, 3'3 diaminobenzidine (DAB) was added to the slides. Slides were visualized by light microscopy and images were acquired with IC Capture 2.2 software and processed with imageJ software version 1.52g.

2.10. Morphometric Analysis of Nerve

The midportion of the graft $n = 6$ from each group was fixed with 2.5% v/v glutaraldehyde (Sigma-Aldrich, Darmstadt, Germany) in 0.1 M phosphate buffer (pH 7.4) for 48 h at room temperature and post-fixed with 1% osmium tetroxide (Sigma-Aldrich, Darmstadt, Germany). The nerve specimens were embedded in epoxy resin, cut into 1-μm, semi-thin sections with an ultramicrotome and stained with 1% toluidine blue (Sigma-Aldrich, Darmstadt, Germany) for light microscopy. Images were digitized with a charge-coupled device camera and analyzed with standard image processing at a magnification of ×1000. Ten random fields from each semi-thin section were analyzed with imaging software (IMARIS 8, Bitplane, Zurich, Switzerland). The sample area was chosen in a systemic, uniform, random manner ensuring that all locations in the nerve cross-section were equally represented. The number of nerve fibers was counted, followed by estimation of mean fiber area and density of myelinated nerve fibers (fibers/μm^2) were determined [7].

2.11. Gastrocnemius Muscle Histology and Muscle Weight Ratio

The gastrocnemius muscle was weighted on an analytical balance immediately after removal from the animals from both sides, normal and experimental, and muscle weight ratio was calculated. Then, the middle part of the muscles was cut and put in a 10% natural formalin solution overnight. The muscles were embedded in paraffin and cut on a microtome into transverse sections at 5 μm, which were subjected to H&E staining followed by observation under light microscope.

2.12. Statistical Analysis

Data was expressed as mean ± standard deviation (SD) and statistical analyses, performed using Graph Pad Prism 6 software (GraphPad Software, San Diego, CA, USA). All data was analyzed with non-parametric student's test, except ADT/ATP essay data that was analyzed with Kruskal–Wallis test and the statically significance level was defined at $p < 0.05$.

3. Results

3.1. Histological Analysis

Histological analysis was performed in order to evaluate the impact of decellularization procedure in hUAs. Specifically, H&E showed the preservation of ECM, while the cellular populations were totally absent. Furthermore, Masson's Trichrome staining revealed the presence of properly oriented collagens in decellularized hUA (Figure 1).

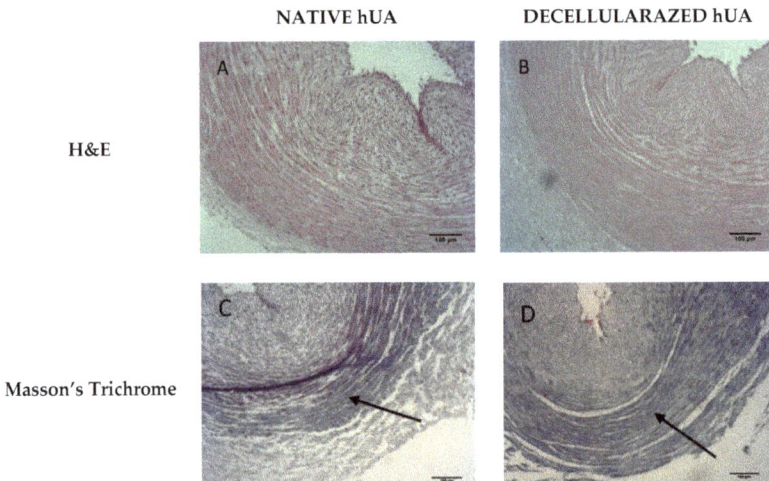

Figure 1. Native and Decellularized human umbilical artery stained with H&E (Hematoxylin and eosin stain), Masson's Trichrome. Native hUA with (**A**) H&E, (**C**) Masson's Trichrome. Decellularized hUA (**B**) H&E, (**D**) Masson's Trichrome. Black arrows were indicated the collagen orientation. Original magnification 10×, scale bars 100 µm.

3.2. Cytotoxicity Tolerance of the Decellularized hUA

Contact cytotoxicity assay results showed that MSCs were expanded and attached successfully to the decellularized hUA segments in 96 well-plates after 48 h of incubation (Figure 2). Moreover, the cells were characterized by the same morphology as the MSCs from negative control group, indicating no cytotoxicity. These findings were further confirmed by determination of ADP/ATP ratio. Specifically, ADP/ATP ratio values were similar between, native, decellularized hUA samples and negative control group. Statistically significant difference was observed only between positive control group, native ($p < 0.001$) and decellularized ($p < 0.001$) hUA samples (Figure 3).

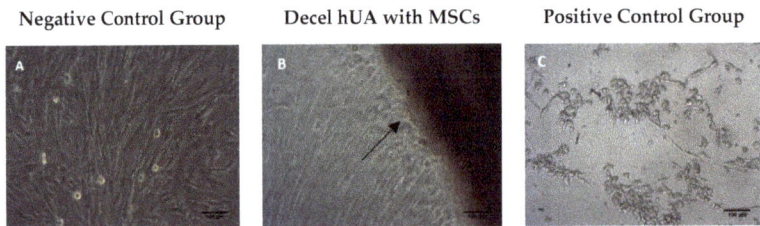

Figure 2. Contact cytotoxicity assay. (**A**) MSCs cultured under normal conditions (negative control group); (**B**) MSCs were co-cultured with decellularized hUA, (**C**) MSCs with SDS (positive control group). Black arrows indicate the contact of MSCs with the segments of decellularized hUA. Original magnification 10×, scale bars 100 µm.

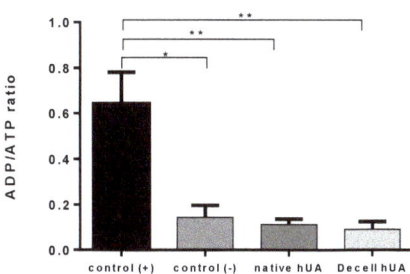

Figure 3. ADP/ATP ratio assay. Low ADP/ATP ratio values indicates that the cells preserved their proliferation capacity. * $p < 0.05$, ** $p < 0.001$.

3.3. Macroscopic Examination of the Experimental Sciatic Nerve In Situ

All animals survived until the end of the experiment and were in good health as indicated by visual inspection. No signs of self-injury were observed in the operated limb. After euthanasia, the implanted nerve conduits derived either from autograft or decellularized group were visually checked. Regenerated nerves were passed through the nerve conduits from both experimental procedures and bridged successfully the 10 mm gap. No evidence of inflammation was observed in both groups. Furthermore, the thickness of the regenerated nerves was similar in animals of both experimental groups (Figure 4).

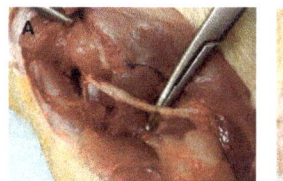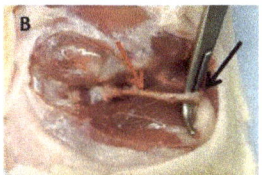

Figure 4. (**A**) autologous nerve graft; (**B**) hUA graft. The red arrow indicates the graft and the black arrow indicates the nerve tissue.

3.4. Motor Function Assessment

The recovery of motor function was assessed by calculating the SFI pre-operative animals, after 4 and 12 weeks. The SFI in all rats prior to surgery was within normal range. SFI values were -10.07 ± 3.38 at the autograft group and -8.26 ± 5.1 at the hUA group without a statistically significant difference. The values of the SFI were reduced at the first post-operative evaluation at week 4 for both groups. SFI values of autograft group and decellularized group was -87.36 ± 6.27 and -80.89 ± 9.22, respectively. No statistically significant difference was observed between the above groups ($p > 0.05$). At 12 weeks, the values of the autograft group showed better improvement than the ones of the hUA group ($p = 0.0013$). Nevertheless, none of these two groups approached the normal values of SFI (Table 1).

Table 1. Values of SFI ± SD.

	Pre-Operative	4 Weeks	12 Weeks
Autograft group	−10.07 ± 3,38	−87.36 ± 6,27	−51.35 ± 7.84
hUA group	−8.26 ± 5,1	−80.89 ± 9,22	−70.56 ± 15.38

3.5. Immunohistochemical Detection of Neurofilaments and Schwann Cells

The transversal section from each group of nerve grafts were stained with anti-NF 200 and anti-S100 antibody for evaluating the axon regeneration. Positive expression of NF200 and S100 detected in all sections (Figure 5A–D).

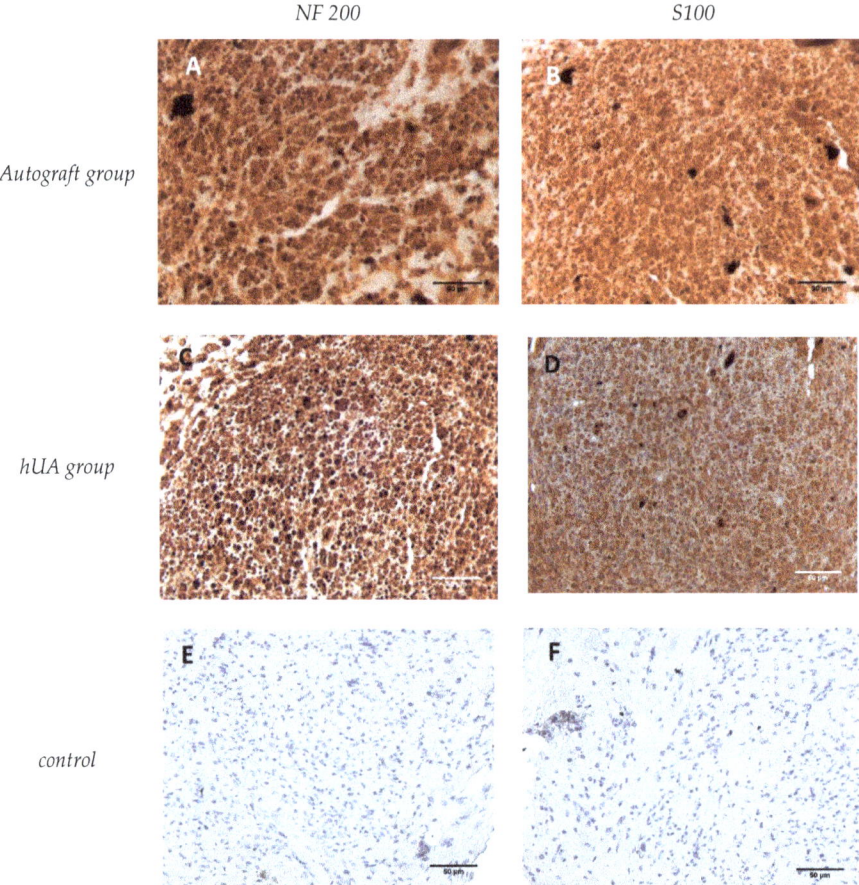

Figure 5. Immunohistochemical results 12 weeks after implantation. (**A,C**) transverse sections of, autograft group and hUA group respectively, stained with NF200 antibody; (**B,D**) transverse sections of autograft group and hUA group respectively, stained with S100; (**E,F**) transverse sections without primary antibody served as a negative control. Original magnification 40× scale bar 50μm.

3.6. Morphometric Analysis

Morphometric analysis was performed in the middle portion of the grafts at week 12. Regenerated myelinated nerve fibers, different in sizes, were observed in each group (Figure 2A,B). The number of nerve fibers between the two groups did not present a statistically significant difference (Figure 6C), but the nerve fibers areas had a wide distribution range (9.87–21.25 μm^2). The area of the hUA group was significantly smaller than that of the autograft group (p = 0.0073, Figure 2D). This trend was also reflected in the density of the nerve fibers (p < 0.0001, Figure 6E).

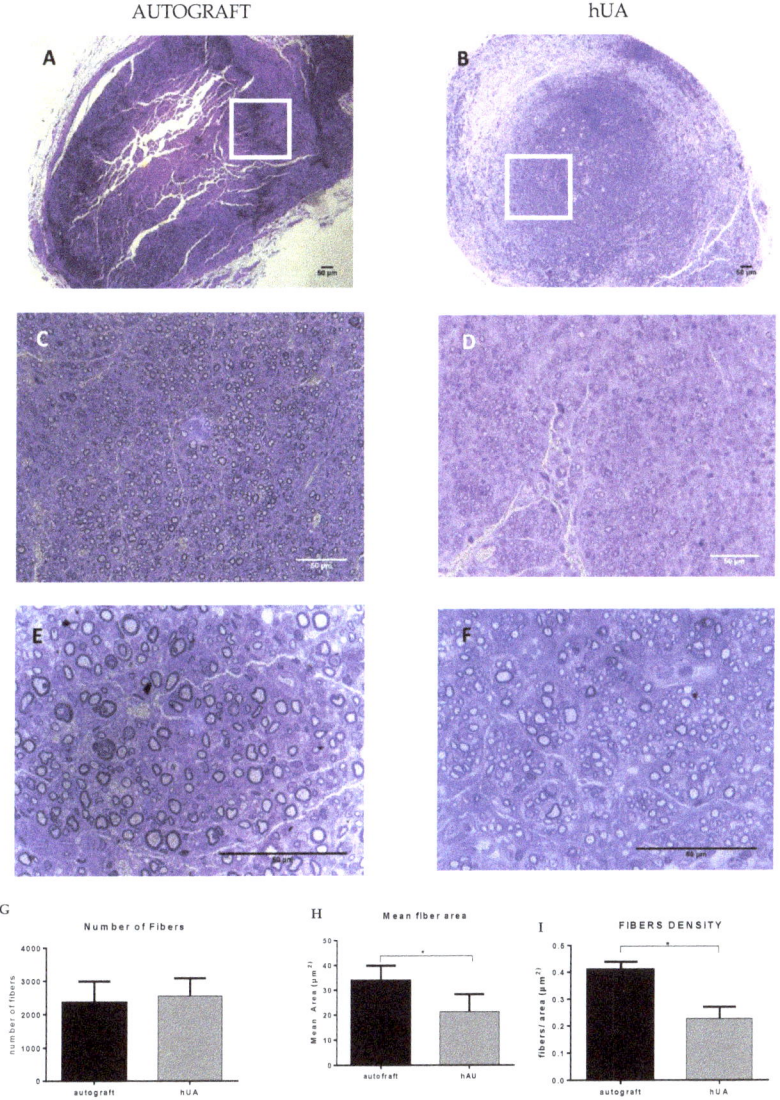

Figure 6. Morphometric analysis of regenerated sciatic nerve (**A,C,F**) toluidine blue stained, middle portion of the autograft after 12 weeks at different magnifications. 10, 40 and 100× respectively; (**B,D,E**) toluidine blue stained, middle portion of the hUA graft after 12 weeks at different magnifications. 10, 40 and 100×, respectively. The white boxes in (**A,B**) highlighted the area that was magnified and presented in the following images (**C,D,E,F**) scale bar 50 µm; (**G**) number of fibers; (**H**) mean fiber area; (**I**) fiber density was evaluated and compared to statistical analysis * $p < 0.05$.

3.7. Gastrocnemius Muscle Histology and Muscle Weight Ratio

In both experimental groups, gastrocnemius muscle showed intense atrophy, while the autograft group presented less atrophy, when compared to the decellularized hUA group. The ratio of muscle mass retention of autologous and decellularized group was 0.55 ± 0.10 and 0.33 ± 0.3, respectively. (Figure 7D). The difference between the two groups is statistically significant ($p < 0.001$).

These results were also supported by the histological evaluation. More specifically, in the autograft group, fibers appeared polygonal with sub-sarcolemmal localization of their nuclei and minimal growth of connective tissue. In the hUA group, the muscle fibers formed small groups of atrophic fibers and more fibrous connective tissues among muscle bundles. In addition, decellularized hUA presented an increased number of cell nuclei when compared to autograft groups (Figure 7A–C).

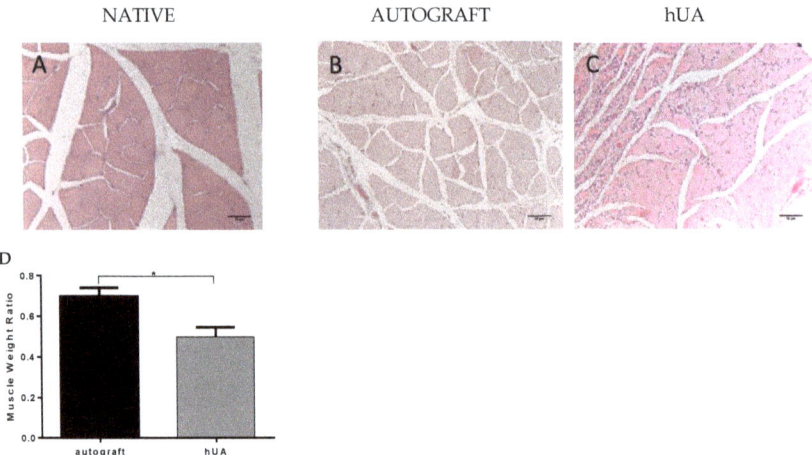

Figure 7. Gastrocnemius muscle histology and weight ratio. (A–C) gastrocnemius muscle transverse section stained with H&E from normal, autograft group and hUA group, respectively. Original magnification 40×; (D) muscle weight ratio was evaluated and compared by statistically analysis * $p < 0.05$.

4. Discussion

Peripheral nerve injuries are very common worldwide, and there is no easily available treatment. Decellularized grafts could be used as an alternative source for nerve conduits. These grafts are characterized by reduced antigenicity and could be a promising therapeutic strategy, when no autologous tissues are available [23].

Different types of decellularized tissues such as nerves and arteries have been used for reconstruction of transected peripheral nerve and showed promising results [24,25]. In this context, hUAs, which can be efficiently isolated from human umbilical cords, a material that is discarded after the gestation, may be good candidates for peripheral nerve reconstruction. The hUA is composed of a complex ECM, which apparently includes collagen, fibronectin, laminin and proteoglycans [17,20]. These proteins, especially laminin, promote neurite and enhance nerve cells adhesion, proliferation and differentiation, thus helping to direct growth cone neurite [26,27]. Due to their importance during the development and the regeneration of the sensory nervous system, laminin, fibronectin and collagen have been successfully used as substrates of tissue culture plastic and poly-3-hydroxybutyrate mats to enhance Schwann cell (SC) response [28]. Furthermore, after the decellularization procedure, the proteoglycans significantly reduced as has been confirmed by others [17]. In this context, chondroitin sulfate, which has a negative impact on nerve guidance and regeneration, can be removed efficiently by the decellularization approach.

Histological analysis indicated the absence of cellular and nuclear residues. Additionally, the structural proteins of the ECM, such as collagen, were preserved when compared to native arteries. After the decellularization of the hUAs, the ATP assay was performed and MSCs were co-cultured within 5 × 5 mm patches of decellularized tissue. As it was expected, the tissue supported the cell attachment and the ADP/ATP assay confirmed the maintenance of proliferation capacity of the cells. MSCs were used for this assay because they are characterized by multilineage differentiation potential. Previous studies have shown that MSCs can differentiate efficiently to neuronal like cells. In addition, future experiments will involve the repopulation of the decellularized hUA with MSCs, implantation in the rat sciatic model, and final evaluation of the function between decellularized and repopulated hUAs. For this purpose, we used MSCs and not neural cells for the cytotoxicity assay. After implantation, artery conduits supported the regeneration sciatic nerves and no inflammatory response was observed.

Previous studies have shown that nerve-conduits have better outcomes, which were similar to autograft results, or even better when they are loaded with different types of cells like adipose-derived stem cells, olfactory cells, Schwann cells, neurotrophic factors or platelet rich plasma [29–34]. In this study, hUA was used alone as an initial step to find out whether it can support elongation of the nerve fibers.

Walk track analysis was performed by estimation of SFI for evaluation of the motor function in animals. Our results demonstrated that neither the autologous nor the decellularized hUA graft restored the SFI close to pre-operative values. Nevertheless, a better functional outcome was observed at the autograft graft group (-51.35). Yeong Kim et al. [32] had similar results on the same week (12th) of SFI evaluation. On the contrary, in other studies, better outcomes were obtained after repairing nerve gaps with autologous grafts (-23.4) [35].

Morphometric and immunohistochemical analysis confirmed the elongation of the nerve fibers and also that the hUA was recellularized and remodeled successfully by the animals.

Immunohistochemical analysis showed positive expression positive of NF200 and S100 in both experimental groups. These findings indicated that decellularized hUA allows the migration of Schwann cells and elongation of the fibers through the umbilical artery tissue.

Morphometric analysis showed the number of fibers between the two experimental groups did not present any statistically significant difference ($p = 0.563$). However, the area and density of the nerve fibers were higher in the autograft group compared to decellularized hUA. These findings may suggest that the nerve fibers at the hUA group were still in a pre-mature stage of [36]. Another parameter to evaluate the re-innervation in the sciatic nerve lesion model is the gastrocnemius muscle weight ratio. When a muscle is denervated, it shifts to degradation, which leads to weight loss [36]. In both experimental groups, the gastrocnemius muscle showed atrophy. The hUA group showed higher atrophy than the autograft group and more cell n. However, normal and smaller muscle fibers co-existed at the hUA group as it was observed at the histological image. However, more cell nuclei were still observed at the hUA group. This can be explained partially by the fact that muscle atrophy was established in decellularized hUA group. Further clarification could be performed by immunohistochemistry for CD11b (macrophage marker) and pro collagen beta 1 (fibroblast markers) [37].

5. Conclusions

In conclusion, this study showed that the decellularized hUA could support the nerve regeneration and could allow the reinnervation of the target organ. Further research in decellularized hUA is needed in order to be used as nerve conduits. Glycosaminoglycans (GAGs) such as chondroitin sulfate, which are important components of hUA, must be properly identified.

In future studies, the decellularized hUAs could be combined with different cell populations or neurotrophic factors, in order to obtain better outcomes, thus bringing them one step closer to clinical application.

Author Contributions: I.G. carried out all the experimental procedures and the statistical analysis of the overall study. M.K. designed and performed the surgical procedure. A.K. and F.A. supervised the surgical procedure. C.S.-G. and E.M. supervised and approved the overall study.

Funding: This research received no external funding.

Conflicts of Interest: The authors declare no conflict of interest.

References

1. Robinson, L.R. Traumatic injury to peripheral nerves. *Muscle Nerve* **2000**, *23*, 863–873. [CrossRef]
2. Taylor, C.A.; Braza, D. The incidence of peripheral nerve injury in extremity trauma. *Am. J. Phys. Med. Rehabil.* **2008**, *8*, 381–385. [CrossRef] [PubMed]
3. Asplund, M.; Nilsson, M. Incidence of traumatic peripheral nerve injuries and amputations in Sweden between 1998 and 2006. *Neuroepidemiology* **2009**, *32*, 217–228. [CrossRef] [PubMed]
4. Ciardelli, G.; Chiono, V. Materials for peripheral nerve regeneration. *Macromol. Biosci.* **2006**, *6*, 13–26. [CrossRef] [PubMed]
5. Alluin, O.; Wittmann, C. Functional recovery after peripheral nerve injury and implantation of a collagen guide. *Biometarials* **2009**, *30*, 363–373. [CrossRef] [PubMed]
6. Daly, W.; Yao, L.A. Biomaterial approach to peripheral nerve regeneration: Bringing the peripheral nerve gap and enhancing functional recovery. *J. R. Soc. Interface* **2012**, *9*, 202–221. [CrossRef] [PubMed]
7. Ansselin, A.D.; Davey, D.F. Axonal regeneration through peripheral nerve grafts: The effect of proximal-distal orientation. *Microsurgery* **1988**, *9*, 103–113. [CrossRef] [PubMed]
8. Nichols, C.M.; Brenner, M.J. Effects of motor versus sensory nerve grafts on peripheral nerve regeneration. *Exp. Neurol.* **2004**, *190*, 347–355. [CrossRef] [PubMed]
9. Konofaos, P.; Ver Halen, J.P. Nerve repair by means of tubulization: Past, present, future. *J. Reconstr. Microsurg.* **2013**, *29*, 149–164. [CrossRef] [PubMed]
10. Crapo, P.M.; Gilbert, T.W. An overview of tissue and whole organ decellularization Processes. *Biomaterials* **2011**, *32*, 3233–3243. [CrossRef] [PubMed]
11. Bellamkonda, R.V. Peripheral nerve regeneration: An opinion of channels scaffolds and anisotropy. *Biometarials* **2006**, *27*, 3515–3518. [CrossRef] [PubMed]
12. Jiang, X.; Lim, S.H. Current applications and future perspectives of artificial nerve conduit. *Exp. Neurol.* **2010**, *223*, 86–101. [CrossRef] [PubMed]
13. Philips, C.; Cornelissen, M. Evaluation methods as quality control in the generation of decellularized peripheral nerve allografts. *J. Neural. Eng.* **2018**, *15*, 021003. [CrossRef] [PubMed]
14. Gu, X.; Ding, F. Construction of tissue engineered nerve grafts and their application in peripheral nerve regeneration. *Prog. Neurobiol.* **2011**, *93*, 204–230. [CrossRef] [PubMed]
15. Karabekmez, F.E.; Duymaz, A. Early clinical outcomes with the use of decellularized nerve allograft for repair of sensory defects within the hand. *HAND* **2009**, *4*, 245–249. [CrossRef] [PubMed]
16. Ahmad, I.; Alhtar, M.S. Use of vein conduit and isolated nerve graft in peripheral nerve repair: A comparative study. *Plast. Surg. Int.* **2014**, *27*, 587968. [CrossRef] [PubMed]
17. Mallis, P.; Gontika, I. Evaluation of decellularization in umbilical cord artery. *Transplant. Proc.* **2014**, *46*, 3232–3239. [CrossRef] [PubMed]
18. Liao, I.C.; Wan, H. Preclinical evaluations of acellular biological conduits for peripheral nerve regeneration. *J. Tissue Eng.* **2013**, *28*. [CrossRef] [PubMed]
19. Jongbae, C.; Jun, H.K. Decellularized sciatic nerve matrix as a biodegradable conduit for peripheral nerve regeneration. *Neural. Regen. Res.* **2018**, *13*, 1796–1803. [CrossRef]
20. Valiyaveettil, M.; Achur, R.N. Characterization of chondroitin sulfate and dermatan sulfate proteoglycans of extracellular matrices of human umbilical cord blood vessels and Wharton's jelly. *Glycoconj. J.* **2004**, *21*, 361–375. [CrossRef] [PubMed]
21. Bain, J.R.; Mackinnon, S.E. Functional evaluation of complete sciatic, peroneal tibial nerve lesions in the rat. *Plast. Reconstr. Surg.* **1989**, *83*, 129–138. [CrossRef] [PubMed]
22. Rafijah, G.; Bowen, A.J. The effects of adjuvant fibrin sealant on the surgical repair of segmental nerve defects in an animal model. *J. Hand Surg. Am.* **2013**, *38*, 847–855. [CrossRef] [PubMed]
23. Gilbert, T.W.; Sellaro, T.L. Decellularization of tissues and organs. *Biomaterials* **2006**, *27*, 3675–3683. [CrossRef] [PubMed]
24. Sun, F.; Zhou, K. Combined use of decellularized allogeneic artery conduits with autologous transdifferentiated adipose-derived stem cells for facial nerve regeneration in rats. *Biomaterials* **2011**, *23*, 8118–8128. [CrossRef] [PubMed]
25. Wakimura, Y.; Wang, W. An experimental study to bridge a nerve gap with a decellularized allogeneic nerve. *Plast. Reconstr. Surg.* **2015**, *136*, 319–327. [CrossRef] [PubMed]

26. Barcelos, A.S.; Rodriges, A.C. Inside out vein graft and inside out artery graft in rat sciatic nerve repair. *Microsurgery* **2003**, *23*, 66–71. [CrossRef] [PubMed]
27. Thanos, P.K.; Okajima, S. Ultrastructure and cellular biology of nerve regeneration. *J. Reconstr. Microsurg.* **1998**, *14*, 423–436. [CrossRef] [PubMed]
28. Armstrong, S.J.; Wiberg, M. ECM molecules mediate both Schwann cell proliferation and activation to enhance neurite outgrowth. *Tissue Eng.* **2007**, *13*, 2863–2870. [CrossRef] [PubMed]
29. Zu, S.; Ge, J. A synthetic oxygen carrier olfactory ensheathing cell composition system for the promotion of sciatic nerve regeneration. *Biometerials* **2014**, *35*, 1450–1461. [CrossRef]
30. Jiang, X.; Lim, S.H. Current application and future perspectives of artificial conduits. *Exp. Neurol.* **2010**, *223*, 86–101. [CrossRef] [PubMed]
31. Zheng, C.; Zhu, Q.C. Improved peripheral nerve regeneration using acellular nerve allografts loaded with platelet-rich plasma. *Tissue Eng. Part A* **2014**, *20*, 3228–3240. [CrossRef] [PubMed]
32. Kim, J.Y.; Jeon, W.J. An iside out vein graft filled with platelet-rich plasma for repair of a short sciatic nerve defect in rats. *Neural Reg Res.* **2014**, *9*, 1351–1357. [CrossRef]
33. Sabongi, R.G.; De Rizzo, L.A. Nerve regeneration: Is there an alternative to nervous graft? *J. Reconstr. Microsurg.* **2014**, *30*, 607–616. [CrossRef] [PubMed]
34. Sarker, M.D.; Saman, N. Regeneration of peripheral nerves by nerve guidance conduits: Influence of design, biopolymers, cells, growth factors, and physical stimuli. *Prog. Neurobiol.* **2018**, in press.. [CrossRef] [PubMed]
35. Komiyama, T.; Nakao, Y. Novel technique for peripheral nerve reconstruction in the absence of an artificial conduit. *J. Neurosci. Methods* **2004**, *134*, 133–140. [CrossRef] [PubMed]
36. Yu, W.; Zhao, W. Sciatic nerve regeneration in rats by a promising electrospun collagen/poly(ε-caprolactone) nerve conduit with tailored degradation rate. *BMC Neurosci.* **2011**, *12*, 68. [CrossRef] [PubMed]
37. Bodine, S.C.; Latres, E. Identification of ubiquitin ligases required for skeletal muscle atrophy. *Science* **2001**, *294*, 1704–1708. [CrossRef] [PubMed]

© 2018 by the authors. Licensee MDPI, Basel, Switzerland. This article is an open access article distributed under the terms and conditions of the Creative Commons Attribution (CC BY) license (http://creativecommons.org/licenses/by/4.0/).

Article

Evaluation of HLA-G Expression in Multipotent Mesenchymal Stromal Cells Derived from Vitrified Wharton's Jelly Tissue

Panagiotis Mallis [1], Dimitra Boulari [1], Efstathios Michalopoulos [1,*], Amalia Dinou [1], Maria Spyropoulou-Vlachou [2] and Catherine Stavropoulos-Giokas [1]

1. Hellenic Cord Blood Bank, Biomedical Research Foundation Academy of Athens, 4 Soranou Ephessiou Street, 115 27 Athens, Greece
2. Immunology Department-Tissue Typing Lab, "Alexandra" General Hospital of Athens, Lourou Street, 11528 Athens, Greece
* Correspondence: smichal@bioacademy.gr; Tel.: +30-210-659-7331

Received: 18 October 2018; Accepted: 31 October 2018; Published: 1 November 2018

Abstract: Background: Mesenchymal Stromal Cells (MSCs) from Wharton's Jelly (WJ) tissue express HLA-G, a molecule which exerts several immunological properties. This study aimed at the evaluation of HLA-G expression in MSCs derived from vitrified WJ tissue. Methods: WJ tissue samples were isolated from human umbilical cords, vitrified with the use of VS55 solution and stored for 1 year at -196 °C. After 1 year of storage, the WJ tissue was thawed and MSCs were isolated. Then, MSCs were expanded until reaching passage 8, followed by estimation of cell number, cell doubling time (CDT), population doubling (PD) and cell viability. In addition, multilineage differentiation, Colony-Forming Units (CFUs) assay and immunophenotypic analyses were performed. HLA-G expression in MSCs derived from vitrified samples was evaluated by immunohistochemistry, RT-PCR/PCR, mixed lymphocyte reaction (MLR) and immunofluorescence. MSCs derived from non-vitrified WJ tissue were used in order to validate the results obtained from the above methods. Results: MSCs were successfully obtained from vitrified WJ tissues retaining their morphological and multilineage differentiation properties. Furthermore, MSCs from vitrified WJ tissues successfully expressed HLA-G. Conclusion: The above results indicated the successful expression of HLA-G by MSCs from vitrified WJ tissues, thus making them ideal candidates for immunomodulation.

Keywords: Mesenchymal Stromal Cells; Wharton's Jelly tissue; HLA-G; mixed lymphocyte reaction; vitrification; VS55; long term storage

1. Introduction

Multipotent Mesenchymal Stromal Cells, also known as Mesenchymal Stem Cells can efficiently be used in a wide variety of tissue engineering and regenerative medicine approaches, such as treatment of bone disorders and regeneration of cardiovascular tissue [1–3]. In addition, MSCs are characterized by critical immunomodulatory properties and could be ideal candidates for the regulation of the immune response [4,5].

According to the International Society for Cellular Therapies (ISCT), MSCs are a fibroblastic cell population, which can be differentiated under defined conditions to mesodermal lineages such as "adipocytes", "osteocytes" and "chondrocytes" [6]. Moreover, human MSCs express specific clusters of differentiation (CDs), including CD73 (ecto-5'-nucleotidase), CD105 (endoglin), CD90 (Thy-1), while lacking totally the expression of CD45 (lymphocyte common antigen), CD34 (hematopoietic stem cell antigen) and HLA class II [6].

MSCs are well known for their immunomodulatory-immunosuppressive properties and their potential use in graft-versus-host disease (GVHD) and autoimmune disorders [7–9].

The immunomodulation, which is induced by MSCs, can be performed either with cell-cell interaction or by secreted factors [10]. A variety of secreted molecules with known immunomodulatory properties including Prostaglandin E2 (PGE2), IL-10, indoleamine 2,3-dioxygenase (IDO) and Human Leukocyte Antigen-G (HLA-G), is being produced by MSCs [10]. Among these factors, HLA-G seems to exert key immunosuppressive properties. HLA-G plays crucial role in preventing the rejection of the semiallogenic fetus by the mother, and also can be used as pre-eclampsia biomarker. HLA-G is non-classical HLA class I molecule, which is located to chromosome 6 (locus p21.1-21.3) in humans. Furthermore, HLA-G is characterized by membrane bound isoforms (HLA-G1-4) and by soluble isoforms (HLA-G5-7). These isoforms can regulate various immune responses such as the inhibition of T cell and natural killer (NK) cell proliferation, as long as the expansion of $CD4^+CD25^+FOXP3^+$ regulatory T cells [11]. The expression of HLA-G in MSCs can be modulated by Interferon-γ (IFN-γ) and IL 10, which can be induced towards allorecognition by various immune cells such as mononuclear and dendritic cells. HLA-G also can be used as a potent marker for MSCs with improved immunosuppressive functions in order to be applied in regenerative medicine and allotransplantation.

Most times, prolonging culture and expansion of MSCs are required for obtaining sufficient cell numbers in order to be used for host immune regulation. By increasing the in vitro cultivation time, this could induce epigenetic modifications and decrease telomere length, which can affect significantly the MSC's characteristics, such as proliferation potential, mesodermal differentiation ability and immunophenotypic properties [12]. In addition, increased cultivation of MSCs may possess a high risk for microbial contamination. Moreover, BM and adipose-derived MSCs require invasive procedures for the primary cell isolation. On the other hand, WJ MSCs have at least similar characteristics with the MSCs from the above sources and exerts the same immunoregulatory properties [13]. In this way, the vitrification and storage of WJ tissue over a long time period may be used as an alternative strategy to obtain MSCs at any desired time point. Cryopreservation by vitrification relies on the use of a combination of high and low molecular weight cryoprotective agents, protecting sufficiently the extracellular matrix (ECM) and tissue resident cells [14]. Vitrification approach reduces the ice crystal formation, thus preserving better the ECM and its mechanical properties. This approach used initially in the storage of human oocytes and embryos and its use has been extended in tissue engineering applications [14]. Several reports, have shown that WJ tissue can be cryopreserved properly followed by efficient isolation and expansion of MSCs, thus decreasing significantly the cultivation period [15,16]. However, thermomechanical stress which is induced by the vitrification and thawing procedures and the long-time storage period may alter the MSC's functional and phernotypic characteristics, including the HLA-G expression. Until now, several research groups have evaluated the immunomodulatory properties of MSCs derived from WJ tissue, BM and adipose tissue [17–19]. However, little is known regarding the expression of HLA-G from MSCs derived from vitrified WJ tissue.

Under this scope, the aim of this study was to evaluate the HLA-G expression in MSCs derived either by vitrified and non-vitrified WJ tissue.

2. Materials and Methods

2.1. Isolation of WJ Tissue Segments

WJ tissue segments ($n = 30$, $l = 7$ cm) were isolated from fresh human umbilical cords that were transferred to Hellenic Cord Blood Bank (HCBB). Human umbilical cords (hUCs, $n = 30$, $l = 10$ cm) were obtained from normal and caesarian deliveries, after signed informed consent by the mothers. The informed consent for this study was in accordance with the declaration of Helsinki and approved by Institution's Bioethics Committee. The hUCs were kept in Phosphate Buffer Saline 1x (PBS 1x, Gibco, Life Technologies, Grand Island, NY, USA) supplemented with 10 U/mL Penicillin and 10 µg/mL

Streptomycin (Gibco, Life Technologies, Grand Island, NY, USA) and processed immediately to WJ tissue isolation. Briefly, the umbilical vessels were discarded and the exposed WJ tissue was isolated and transferred to 15 mL polypropylene falcon tubes (BD Biosciences Bedford, Bedford, MA, USA) with PBS 1x (Gibco, Life Technologies, Grand Island, NY, USA) until further use.

2.2. Vitrification of WJ Tissue

Isolated WJ tissue ($n = 10$, $l = 2$ cm), were cut into segments using sterile instruments. Specifically, each WJ tissue was divided into 3 segments with an average length of 2 cm. A number of 10 samples of WJ tissue was placed into cryotubes (BD Biosciences Bedford, Bedford, MA, USA) with approximately 2 mL of precooled VS55 vitrification solution. VS55 solution was consisted of 3.10 M DMSO, 3.10 M formamide, 2.21 M 1,2-propanediol (Sigma Aldrich, St. Louis, MO, USA) in Euro-Collins solution (IndiaMART, Noida, India).

The cryotubes contained the WJ tissue samples were rapidly cooled (43 °C/min) until reached -100 °C, followed by slow cooling (3 °C/min) to -135 °C. Finally, the samples were transferred to liquid nitrogen at -196 °C. The samples were stored in this state for a time period of 1 year. The same procedure was performed in WJ tissue samples ($n = 10$, $l = 2$ cm), without the addition of any cryoprotective agent. These tissue segments were served as positive control group and will be referred as CPA-free samples. Non-vitrified fresh WJ tissue samples ($n = 10$, $l = 2$ cm) were also used and served as negative control group for this study.

2.3. Thawing of WJ Tissue

After 1 year of storage in liquid nitrogen, vitrified ($n = 10$, $l = 2$ cm) and CPA-free ($n = 10$, $l = 2$ cm) WJ tissue samples were thawed. Briefly, the cryotubes were quickly transferred from -196 °C to waterbath at 37 °C. Then, each sample was transferred to 50 mL polypropylene falcon tubes (BD Biosciences Bedford, Bedford, MA, USA) with 40 mL of PBS 1x (Gibco, Life Technologies, Grand Island, NY, USA) and centrifuged at $500\times g$ for 6 min. Finally, the supernatant was discarded and the WJ tissue samples were placed to 100 mm^2 Petri dish (ThermoFisher Scientific, Waltham, MA, USA) in order to proceed to isolation of WJ-MSCs.

2.4. Isolation and Expansion of WJ-MSCs

WJ tissue derived either from non-vitrified ($n = 10$, $l = 2$ cm), vitrified ($n = 10$, $l = 2$ cm) and CPA-free ($n = 10$, $l = 2$ cm) samples were trimmed with the use of sterile instruments and then each sample was placed separately in 6-well plate (Costar, Corning Life, Canton, MA, USA). Finally, 1 mL of standard culture medium was added in each well, and the 6-well plates were remained in humidified atmosphere with 5% CO_2 at 37 °C for a total time period of 18 days. When confluency observed, the cells were detached using 0.25% trypsin-EDTA solution (Gibco, Life Technologies, Grand Island, NY, USA) and transferred to 75 cm^2 cell culture flask (Costar, Corning Life, Canton, MA, USA). The cells remained in 75 cm^2 cell culture flask (Costar, Corning Life, Canton, MA, USA) for additional 10 days, upon reaching confluency. Then, the cells were trypsinized and transferred to 175 cm^2 cell culture flask (Costar, Corning Life, Canton, MA, USA). The same procedure was performed until the cells reached passage (P) 8. The standard culture medium used in this study, consisted of α-Minimum Essentials Medium (α-MEM, Gibco, Life Technologies, Grand Island, NY, USA) supplemented with 15% v/v fetal bovine serum (FBS, Gibco, Life Technologies, Grand Island, NY, USA) and 1% v/v penicillin (Gibco, Life Technologies, Grand Island, NY, USA) and 1% v/v streptomycin (Gibco, Life Technologies, Grand Island, NY, USA).

2.5. Histological Analysis of WJ Tissue

Histological analysis of non-vitrified ($n = 5$), vitrified ($n = 5$) and CPA-free ($n = 5$) WJ tissue samples with Hematoxylin and Eosin (H&E, Sigma-Aldrich, Darmstadt, Germany) stain, was performed. Briefly, the WJ tissue samples were fixed with 10% v/v neutral formalin buffer (Sigma-Aldrich,

Darmstadt, Germany), dehydrated, paraffin embedded and sectioned at 5 µm. Then, the slides were rehydrated and stained with H&E stain. Finally, images were acquired with Leica DM LS2 (Leica, Microsystems, Wetzlar, Germany) microscope and processed with IC Capture v 2.4 software (Imaging Source, Bremen, Germany).

2.6. Multi-Differentiation Capacity of WJ-MSCs

The differentiation ability of WJ-MSCs towards "osteogenic", "adipogenic" and "chondrogenic" lineages was assessed. For this purpose, WJ-MSCs P3 from non-vitrified ($n = 3$) and vitrified ($n = 3$) tissue samples were used. Specifically, WJ-MSCs at a density of 5×10^4 cells were plated in each well of 6-well plates (Costar, Corning Life, Canton, MA, USA) with standard culture medium for "osteogenic" and "adipogenic" differentiation. When, the cells reached 80% of confluency, the culture medium was aspirated and briefly washes with PBS 1x (Gibco, Life Technologies, Grand Island, NY, USA) were performed. Then, PBS 1x was removed totally and the cells were subjected to differentiation.

"Osteogenic" differentiation was performed by addition of basal medium (Mesencult, StemCell Technologies, Vancouver, BC, Canada) supplemented with 15% v/v Osteogenic stimulatory supplements (StemCell technologies, Vancouver, BC, Canada), 0.01 mM dexamethasone (StemCell technologies, Vancouver, BC, Canada) and 50 ng/mL ascorbic acid (StemCell technologies, Vancouver, BC, Canada). The total time period needed for the differentiation to "osteocytes" was 25 days and then Alizarin Red-S (Sigma-Aldrich, Darmstadt, Germany) staining was performed in order to confirm the successful differentiation. WJ-MSCs were subjected to "adipogenic" differentiation by using the basal medium (Mesencult, StemCell Technologies, Vancouver, BC, Canada) supplemented with 10% v/v of adipogenic stimulatory supplements (StemCell Technologies, Vancouver, BC, Canada). After 25 days of culture, Oil Red-O (Sigma-Aldrich, Darmstadt, Germany) staining was performed.

"Chondrogenic" differentiation was conducted in 3D spheroid cultures, by transferring WJ-MSCs at a density of 35×10^4 cells in 15 mL polypropylene falcon tubes (BD Biosciences Bedford, USA). "Chondrogenic" differentiation medium consisted of high glucose D-MEM (Sigma-Aldrich, Darmstadt, Germany) supplemented with 0.01mM dexamethasone (StemCell technologies, Vancouver, BC, Canada), 40 µg/mL ascorbic acid-2 phosphate (StemCell Technologies, Vancouver, BC, Canada), 10 ng/mL transforming growth factor-β1 (TGF-β1, Sigma-Aldrich, Darmstadt, Germany), and 100 µL of insulin-transferin selenium liquid medium 100× (ITS 100×, StemCell technologies, Vancouver, BC, Canada). After 30 days of culture, the pellets were fixed with 10% v/v neutral formalin buffer (Sigma-Aldrich, Darmstadt, Germany), dehydrated, paraffin embedded and sectioned at 5 µm. Alcian blue (Sigma-Aldrich, Darmstadt, Germany) was performed in each sample for the determination of "chondrogenic" differentiation. Images were acquired with Leica DM LS2 (Leica, Microsystems, Wetzlar, Germany) microscope and processed with IC Capture v 2.4 software (Imaging Source, Bremen, Germany).

2.7. Colony-Forming Unit-Fibroblast (CFU-F) Assay in WJ-MSCs

WJ-MSCs derived from non-vitrified ($n = 3$) and vitrified ($n = 3$) tissue samples were seeded at a density of 500 cells/well on 6-well plates (Costar, Corning Life, Canton, MA, USA), followed by addition of 1 mL of standard culture medium. The cultures remained in a humidified atmosphere with 5% CO_2 at 37 °C for 15 days. The culture medium was changed twice a week. After 15 days of cultivation, WJ-MSCs were fixed with 10% v/v neutral formalin buffer (Sigma-Aldrich, Darmstadt, Germany) for 5 min. Finally, Giemsa (Sigma-Aldrich, Darmstadt, Germany) staining and manual counting of the colonies by two independent observers were performed. Images were acquired with Leica DM LS2 (Leica, Microsystems, Wetzlar, Germany) microscope and processed with IC Capture v 2.2 software (Imaging Source, Bremen, Germany).

2.8. Cell Doubling Time, Population Doubling and Cell Viability Estimation

Total cell number, cell doubling time (CDT), population doubling (PD) and cell viability was measured after each passage of WJ-MSCs until reached passage 8. Initially, WJ-MSCs at a density of 2×10^5 cells were placed in 75 cm^2 cell culture flasks (Costar, Corning Life, Canton, MA, USA). The CDT was estimated based on the following equation

$$CDT = \frac{\log_{10}(N/N0)}{\log_{10}(2)} \times (T)$$

The determination of PD rate was performed according to the equation

$$PD = \frac{\log_{10}(N/N0)}{\log_{10}(2)}$$

where N was the number of cells at the end of the culture, $N0$ was the number of seeded cells and T was the culture duration in hours.

The viability of WJ-MSCs was performed using the Trypan blue (Sigma Aldrich, St. Louis, MO, USA). Total cell number was microscopically counted in Neubauer slide (Celeromics, Valencia, Spain). The cell viability and total cell number counting were performed by two independent observers. In addition, WJ-MSCs derived only from non-vitrified (n = 5) and vitrified (n = 5) tissue samples were used for the above measurements.

2.9. Flow Cytometric Analysis

WJ-MSCs at passage 3 derived from vitrified (n = 3) and non-vitrified (n = 3) tissue samples, were analyzed with flow cytometry for the expression of specific CDs. WJ-MSCs were tested for CD90 (Thy-1), CD105 (endoglin), CD73 (ecto-5'-nucleotidase), CD29 (integrin subunit), CD19 (pan-B-cell marker), CD31 (pan-endothelial cell marker), CD45 (pan-hematopoietic cell marker), CD34 (hematopoietic stem cell marker), CD14 (TLR-4 co-receptor), CD3 (T-cell co-receptor), CD19 (B-Cell marker), HLA-DR (HLA class II antigen) and HLA-ABC (HLA class I antigen). In addition, anti-CD90, HLA-ABC, CD29, CD19, CD31 and CD45 were conjugated with fluorescein isothiocyanate and anti-CD105, CD73, CD44, CD3, CD34, CD14 and HLA-DR were conjugated with phycoerythrin. All monoclonal antibodies used for this assay, were purchased from Immunotech (Immunotech, Beckman Coulter, Marseille, France). The flow cytometric analysis was performed in Cytomics FC 500 (Beckman Coulter, Marseille, France) coupled with the CXP Analysis software (Beckman Coulter, Marseille, France).

2.10. Evaluation of HLA-G Expression

HLA-G expression was evaluated in WJ-MSCs P3 derived from non-vitrified (n = 5) and vitrified (n = 5) WJ tissue samples. Total mRNA was isolated using the TRI-reagent (Sigma-Aldrich, Darmstadt, Germany) according to manufacturer's instructions. Reverse transcription polymerase chain reaction (RT-PCR) with Omniscript RT Kit (Qiagen, Hilden, Germany) using 800 ng of RNA was performed. Complementary DNA (cDNA) was amplified with PCR by using specific primers (Table 1). Taq PCR Master Mix (Qiagen, Hilden, Germany) was applied according to manufacturer's instructions. PCR was performed on Eppendorf Master Cycler (Eppendorf, Hamburg, Germany) and involved the following steps: initial denaturation at 95 °C for 15, denaturation at 94 °C for 30 s, annealing at 60–61 °C for 90 s and final extension at 72 °C for 3 min. The total number of cycles used in this study was 35. Finally, the PCR products were analyzed on 1% w/v agarose gel (Sigma-Aldrich, Darmstadt, Germany).

Table 1. Primer Set used in RT-PCR.

Gene	Forward Sequence	Target Region	Reverse Sequence	Target Region	Amplicon Size
HLA-G1	AGGAGACACGGAACACCAAG	Exon 2	CCAGCAACGATACCCATGAT	Exon 5	685
HLA-G5	AACCCTCTTCCTGCTGCTCT	Exon 1	GCCTCCATCTCCCTCCTTAC	Intron 4	895
HLA-G7	AACCCTCTTCCTGCTGCTCT	Exon 1	TTACTCACTGGCCTCGCTCT	Intron 2	331
GAPDH	AAGGGCCCTGACAACTCTTT	-	CTCCCCTCTTCAAGGGGTCT	-	244

2.11. Flow Cytometric Analysis of HLA-G

HLA-G expression in WJ-MSCs P3 was determined by flow cytometric analysis. Briefly, MSCs derived from non-vitrified (n = 3) and vitrified WJ (n = 3) tissue samples at a density of 10^4 cells were placed in flow cytrometric tube. Initially, incubation with monoclonal antibody against human HLA-G (1:1000, Catalog MA1-10359, ThermoFisher Scientific, Waltham, MA, USA) was performed for 60 min. Then, 1 mL of PBS 1x was added and the samples were centrifuged at 500× g for 6 min. The supernatant was discarded, followed by addition of secondary FITC-conjugated mouse IgG antibody (1:100, Sigma-Aldrich, Darmstadt, Germany). The samples were incubated for 30 min at room temperature and then centrifugated at 500× g for 6 min. Finally, the samples were analyzed with Cytomics FC 500 (Beckman Coulter, Marseille, France) coupled with the CXP Analysis software (Beckman Coulter, Marseille, France).

2.12. Immunohistochemistry for HLA-G Determination

Immunohistochemistry against HLA-G was applied in non-vitrified (n = 5) and vitrified (n = 5) WJ tissue samples. Vitrified WJ tissue samples were thawed and remained with standard culture medium for 48 h in humidified atmosphere at 37 °C prior to the performance of the immunohistochemistry assay. The EnVision FLEX Mini kit, high pH (Agilent Technologies, Santa Clara, CA, USA) was used according to manufacturer's instructions. Briefly, the tissues were cryoembedded and sectioned at 10 μm. Initially, endogenous peroxidase blocking, followed by addition of anti-HLA-G (1:1000, Catalog MA1-10359, ThermoFisher Scientific, Waltham, MA, USA), was performed. Briefly washes were performed, followed by addition of mouse secondary IgG antibody. After, 30 mins of incubation, DAB was added. Hematoxylin (Sigma-Aldrich, Darmstadt, Germany) staining was performed. Finally, the slides were dehydrated and mounted. Images were acquired with Leica DM LS2 (Leica, Microsystems, Wetzlar, Germany) microscope and processed with IC Capture v 2.4 software (Imagingsource, Bremen, Germany).

2.13. Mixed Lymphocyte Reaction (MLR)

Isolation of peripheral blood mononuclear cells (PBMNCs) was performed from two volunteers after signed informed consent. The informed consent was in accordance with the declaration of Helsinki and has been approved by the bioethics committee of BRFAA. The MLR was performed according to a previous published protocol [20]. After blood sampling, gradient centrifugation was performed with the use of Ficoll Solution (Sigma-Aldrich, Darmstadt, Germany) at 500× g for 30 min. Then, the PBMNCs from the first volunteer (stimulator cells) were treated with 25 μg/mL mitomycin (Sigma-Aldrich, Darmstadt, Germany) for 30 min at 37 °C. The PBMNCs from the second volunteer (responder cells) were used without any treatment for the MLR assay. Total cell number were estimated by two different observers using Trypan blue (Sigma-Aldrich, Darmstadt, Germany). WJ-MSCs P3 obtained from non-vitrified (n = 10) and vitrified (n = 10), were plated at a density of 1×10^4 cells on U-bottom 96-well plates (Costar, Corning Life, Canton, MA, USA) with 200 μL of standard culture medium. Equal number of responder and stimulator PBMNCs were added to each well of 96 well plate (Costar, Corning Life, Canton, MA). Cultivation of 96 well plates was performed for 5 days at 37 °C in 5% CO_2. Finally, the responder PBMNCs were counted with MTT cell growth assay kit (Sigma-Aldrich, Darmstadt, Germany).

Specifically, MLR assay involved the following interactions between cellular populations. Stimulator cells ($n = 10$) without the addition of any other cell population, responder cells ($n = 10$) without the addition of any other cell population, interaction between stimulator ($n = 10$) and responder ($n = 10$) cells which will be referred as MLR, MLR with MSCS derived either from non-vitrified ($n = 10$) or vitrified ($n = 10$) WJ tissue samples, stimulator cells ($n = 10$) with MSCs from both experimental procedures ($n = 20$) and responder cells ($n = 10$) with MSCs from both experimental procedures ($n = 10$).

2.14. Indirect Immunofluorescence for HLA-G Determination

Indirect immunofluorescence against HLA-G in WJ-MSCs obtained from non-vitrified ($n = 5$) and vitrified ($n = 5$) WJ tissue samples was performed. An average of 1×10^4 WJ-MSCs were seeded on culture slides (Sigma-Aldrich, Darmstadt, Germany), followed by addition of 1 mL of standard culture medium. After 10 days, the culture slides were microscopically checked and when confluency observed, the indirect immunofluorescence was performed. For this purpose, the WJ-MSCs were fixed for 10 min with 10% v/v neutral formalin buffer (Sigma-Aldrich, Darmstadt, Germany). Antigen retrieval and blocking of cells was applied in all samples, followed by addition of monoclonal antibody against human HLA-G (1:1000, Catalog MA1-10359, ThermoFisher Scientific, Waltham, MA, USA). Secondary FITC-conjugated mouse IgG antibody (1:100, Sigma-Aldrich, Darmstadt, Germany) was added. Cell nuclei became evident with DAPI staining (ThermoFisher Scientific, Waltham, MA, USA). Finally, the slides were glycerol mounted and observed under fluorescent microscope. Images were acquired with LEICA SP5 II microscope equipped with LAS Suite v2 software (Leica, Microsystems, Wetzlar, Germany).

2.15. Statistical Analysis

Statistical analysis was performed with GraphPad Prism v 6.01 (GraphPad Software, San Diego, CA, USA). Comparison in cell, number, viability, CDT, PD and MTT cell growth assay between the experimental conditions was performed with the unpaired nonparametric Mann–Whitney U test. Statistically significant difference between group values was considered when p value was less than 0.05. Indicated values are mean ± standard deviation.

3. Results

3.1. Isolation of MSCs from Vitrified WJ-Tissue

MSCs were successfully isolated from fresh non-vitrified and vitrified WJ tissue samples. First evidence of cells from fresh non-vitrified and vitrified WJ tissue samples was observed at day 6 and day 7, respectively (Figure S1). Furthermore, the cells were expanded and confluency observed at day 18 (Figure 1). Then, passage of WJ-MSCs was performed. Specifically, 80% of MSCs from non-vitrified WJ-tissue samples were passaged at 75 cm^2 cell culture flasks and 20% of MSCs were passaged at 25 cm^2 flasks. In regards to MSCs obtained from vitrified WJ-tissue samples, the 70% were passaged at 75 cm^2 flasks and 30% were passaged at 25 cm^2 flasks (Figure S2).

WJ-MSCs from both experimental conditions were spindle-shaped and their morphology retained without any alteration until reaching passage 8 (Figure S3). On the other hand, no cells were isolated from CPA-free WJ tissue samples after 18 days of culture. The cultures remained for additional 10 days in a humidified atmosphere at 37 °C. No viable cells were obtained from CPA-free samples. Under this scope, for the next set of experiments only MSCs obtained from non-vitrified and vitrified WJ tissues were used.

3.2. Histological Analysis

Histological analysis was performed in order to determine the effect of low temperatures in WJ-tissue ultrastructure. Non-vitrified and vitrified WJ tissues were characterized by a dense gelatinous extracellular matrix, rich in resident cellular populations (Figure 2). Moreover, vitrified WJ-tissue appeared to be preserved properly with the current vitrification protocol; thus, no damage occurred in the resident cells. On the other hand, CPA-free WJ-tissue was characterized by extensive damage of the ECM, due to ice crystal formation during the storage procedure (Figure 2). The extensive WJ-tissue damage that was observed in CPA-free samples, could negatively affect the viability of WJ-MSCs.

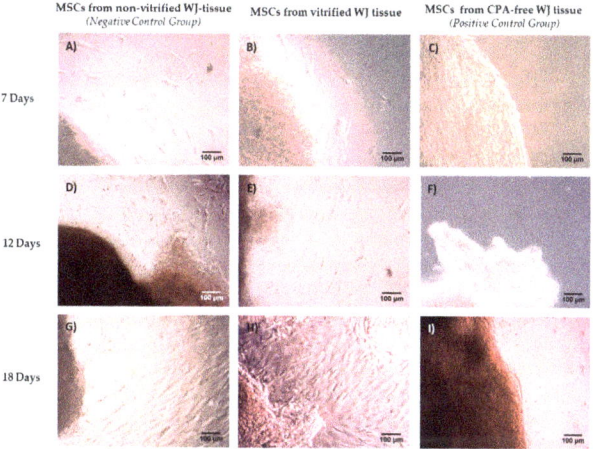

Figure 1. Isolation of MSCs derived from non-vitrified, vitrified and CPA-free WJ tissue. MSCs derived from fresh non-vitrified WJ-tissue (n = 10) at day 7 (**A**), 12 (**D**) and 18 (**G**). MSCs derived from vitrified WJ tissue (n = 10) at day 7 (**B**), 12 (**E**) and 18 (**H**). No MSCs were able to be isolated from CPA-free (n = 10) WJ tissue at day 7 (**C**), 12 (**F**) and 18 (**I**). Original magnification 10×, scale bars 100 µm.

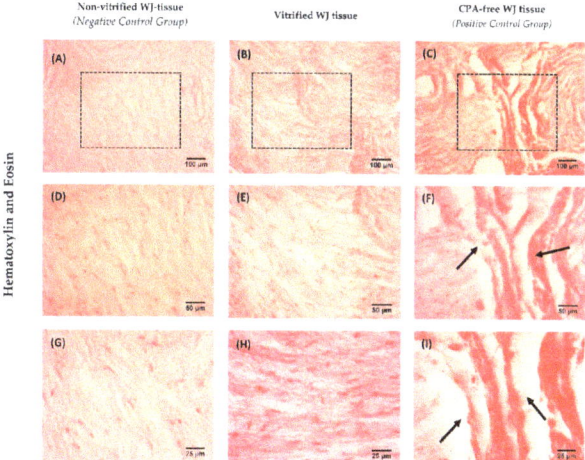

Figure 2. Histological analysis of WJ tissue. Non-vitrified (n = 5) WJ tissue (**A,D,G**), vitrified (n = 5) WJ tissue (**B,E,H**) and CPA-free (n = 5) WJ tissue (**C,F,I**) stained with H&E. The black boxes in 10× images (**A–C**) highlights the region that appeared in higher magnification of 20× and 40×. Black arrows in CPA-free WJ tissue with magnification 20× and 40× indicated the extensive damage of the tissue. Images **A–C**, were acquired with original magnification 10×, scale bars 100 µm. Images **D–F**, were acquired with original magnification 20×, scale bars 50 µm. Images **G–I**, were acquired with original magnification 40×, scale bars 25 µm.

3.3. Characteristics and Growth Kinetics of WJ-MSCs

WJ-MSCs from both experimental conditions exhibited multipotent differentiation potential towards "osteogenic", "adipogenic" and "chondrogenic" lineages as has been confirmed by Alizarin Red S, Oil red O and Alcian blue staining, respectively (Figure 3). Specifically, under "osteogenic" differentiation conditions, WJ-MSCs derived either from non-vitrified and vitrified WJ tissue samples were characterized by calcium deposits, which were visible by Alizarin Red S staining (Figure 3A). In addition, differentiated WJ-MSCs towards "adipogenic" lineages characterized by intracellular lipid droplets, which were stained positive with Oil Red-O staining (Figure 3A). Furthermore, WJ-MSCs differentiated successfully to chondrocytes and stained positive with Alcian blue staining (Figure 3A).

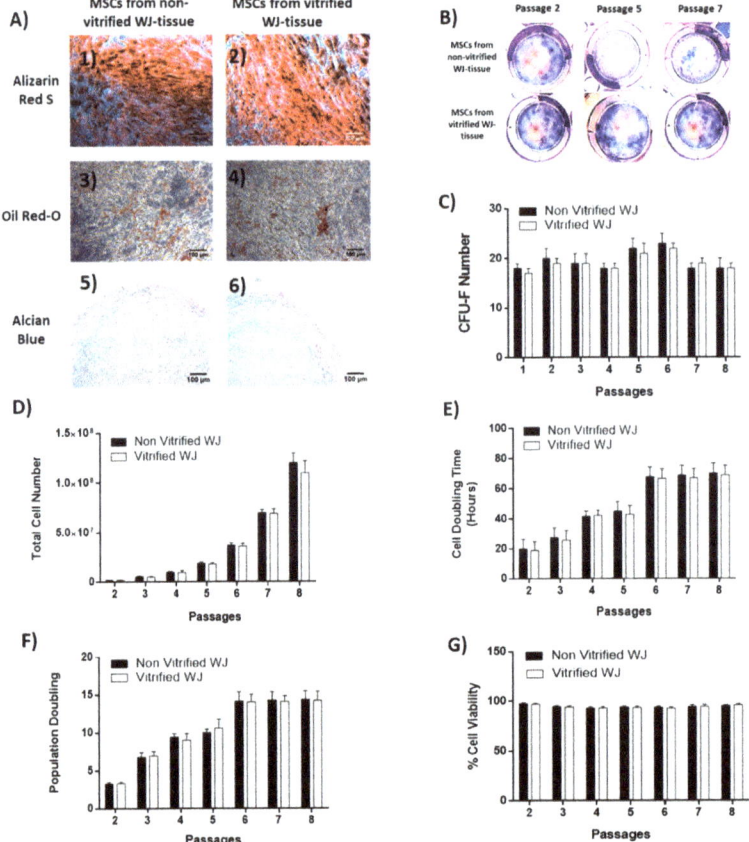

Figure 3. Characteristics of MSCs derived from non-vitrified and vitrified WJ tissue samples. (**A**) Differentiation potential of MSCs from non-vitrified ($n = 3$) WJ tissue samples towards "osteogenic (1)", "adipogenic (3)" and "chondrogenic (5)" lineages. Differentiation potential of MSCs from vitrified ($n = 3$) WJ tissue samples towards "osteogenic (2)", "adipogenic (4)" and "chondrogenic (6)" lineages. Original magnification 10×, scale bars 100 μm (**B**) CFU-F assay of WJ-MSCS obtained from non-vitrified ($n = 3$) and vitrified ($n = 3$) WJ tissue samples. (**C**) Counting of CFUs. Total cell number (**D**), CDT (**E**), PD (**F**), % cell viability (**G**) of MSCs until reaching passage 8 from non-vitrified ($n = 3$) and vitrified ($n = 3$) WJ tissue samples.

A CFU-F assay was performed in order to determine the clonogenic potential of WJ-MSCs. Specifically, MSCs derived from non-vitrified WJ tissue samples presented a higher number of CFU-F when compared to MSCs derived from vitrified tissue, but this increase was not statistically significant (Figure 3B,C). The higher number of CFU-F of WJ MSCs from both experimental conditions was at passage 6 (23 ± 2 and 22 ± 2 CFUs).

In addition, total cell number, CDT, PD and cell viability were calculated in order to determine the characteristics of WJ-MSCs. The total cell number of WJ-MSCs when reaching passage 8 surpassed the 1.1×10^8 cells (Figure 3D). The mean CDT of MSCs derived from non-vitrified and vitrified WJ tissue samples ranged between 20 ± 7 to 70 ± 7 and 19 ± 6 to 69 ± 7 h. In addition, the mean PD of MSCs from the first experimental condition ranged between 3 ± 1 to 14 ± 1 and from the second experimental condition ranged between 3 ± 1 to 14 ± 2. The mean cell viability of MSCs derived from both experimental conditions until reaching passage 8 ranged between 94 ± 1% to 95 ± 1% and 95 ± 1% to 96 ± 1%.

3.4. Immunophenotypic Analysis of WJ MSCs with Flow Cytometer

WJ-MSCs from both experimental conditions expressed positively CD73, CD90, CD105, HLA-ABC, CD29 and CD44 (Table 2). In addition, WJ-MSCs derived either from non-vitrified or vitrified WJ tissue samples were characterized by negative expression of CD3, CD19, CD31, HLA-DR, CD34 and CD45 (Table 2). There were no statistically significant differences in surface markers between WJ-MSCs from both experimental conditions. A detailed list of surface marker expression is shown in Table 2.

Table 2. Flow cytometric analysis of MSCs derived from non-vitrified (n = 3) and vitrified (n = 3) WJ-tissue samples.

Clusters of Differentiation	MSCs Derived from Non-Vitrified WJ Tissue (% Expression)	MSCs Derived from Vitrified WJ Tissue (% Expression)	p-Value
CD73	97.3 ± 0.9	97.0 ± 0.8	0.7431
CD90	96.0 ± 0.6	96.7 ± 0.5	0.6338
CD105	96.3 ± 0.7	96.7 ± 0.5	0.5897
HLA-ABC	97.0 ± 0.1	96.4 ± 2.3	0.7369
CD29	96.0 ± 1.3	95.2 ± 1.5	0.6306
CD44	96.5 ± 0.6	95.2 ± 0.8	0.1489
CD3	1.6 ± 0.2	1.6 ± 0.1	0.9989
CD19	1.3 ± 0.1	1.3 ± 0.5	0.2663
CD31	1.4 ± 0.1	1.4 ± 0.1	0.9978
CD14	1.1 ± 0.2	1.1 ± 0.1	0.8419
HLA-DR	1.1 ± 0.1	1.1 ± 0.1	0.6033
CD-34	1.5 ± 1.1	1.6 ± 0.2	0.4501
CD45	1.4 ± 0.3	1.3 ± 0.1	0.7445

3.5. Immunohistochemistry of HLA-G Expression

Non-vitrified and vitrified WJ tissues successfully expressed the HLA-G, as indicated by immunohistochemistry results. Specifically, vitrified WJ tissue was characterized by the expression of HLA-G, even after 1 year of storage (Figure 4). In addition, non-vitrified and vitrified WJ tissues stained positive for HLA-G in a similar way as the positive control group (Figure S5).

3.6. HLA-G Expression in WJ-MSCs

WJ-MSCs from both experimental conditions were characterized by the expression of *HLA-G*. Specifically, MSCs at passage 3 derived from non-vitrified and vitrified WJ tissue samples successfully expressed the intracellular isoform, *HLA-G1* and the soluble isoforms *HLA-G5* and *HLA-G7* (Figure 5A). Flow cytometric analysis indicated that 96 ± 1% of MSCs obtained from non-vitrified WJ tissue and 95 ± 2% of MSCs from vitrified WJ tissue, expressed the HLA-G (Figure 5B and Table S1).

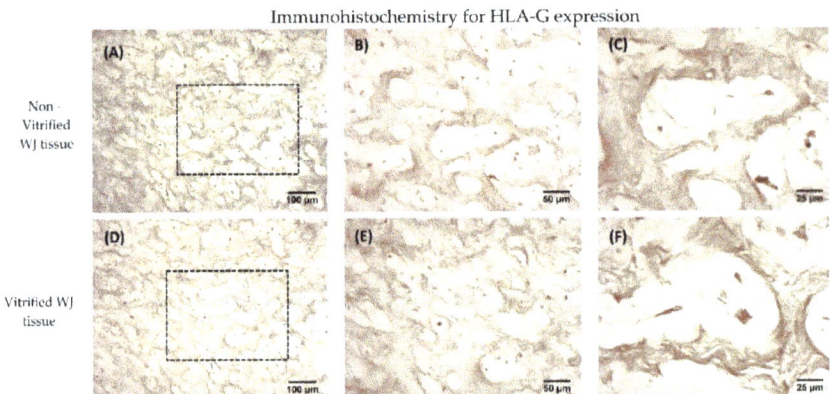

Figure 4. HLA-G expression in WJ tissue. Immunohistochemistry results regarding the HLA-G expression in non-vitrified ($n = 5$) WJ tissue (**A–C**). Immunohistochemistry results regarding the HLA-G expression in vitrified ($n = 5$) WJ tissue (**D–F**). The black boxes in 10× images (**A,D**) highlights the region that was magnified in images **B,C,E,F**. Images **A,B** were acquired with original magnification 10×, scale bars 100 μm. Images **B,E** were acquired with original magnification 20×, scale bars 50 μm. Images **C,F** were acquired with original magnification 40×, scale bars 25 μm.

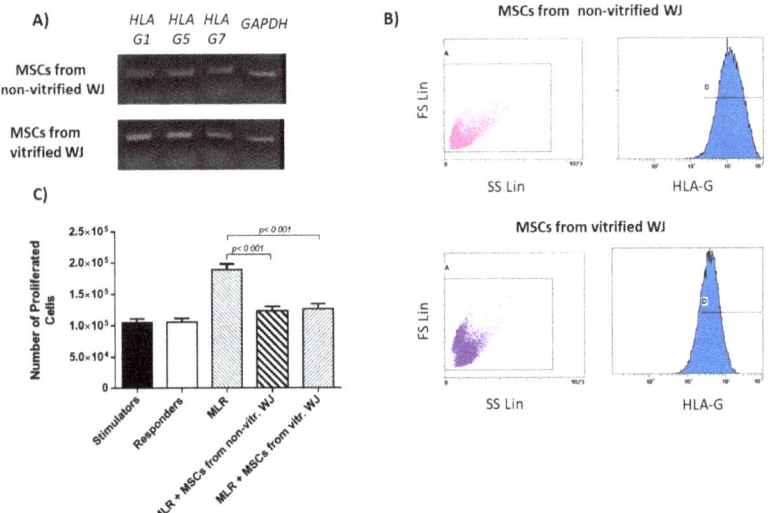

Figure 5. HLA-G expression by WJ-MSCs. (**A**) HLA-G expression of MSCs derived from non-vitrified ($n = 5$) and vitrified ($n = 5$) WJ tissue; (**B**) Characterization of HLA-G expression in MSCs derived from non-vitrified ($n = 3$) and vitrified ($n = 3$) WJ tissue, by flow cytometric analysis. (**C**) Mixed lymphocyte reaction using WJ-MSCs from both experimental conditions. Statistically significant differences were observed between MLR and MLR coupled with MSCs from non-vitrified WJ tissue ($p < 0.001$) and MSCs from vitrified WJ tissue ($p < 0.001$).

MLR results showed that WJ-MSCs from both experimental conditions successfully suppressed the proliferation of PBMNCs as indicated by the mean number of proliferated cells (12×10^4 in both experimental conditions). On the other hand, responder cells presented high proliferation (19×10^4 cells) without the addition of WJ-MSCs (Table S2 and Figure S4). Statistically significant differences between MLR and MLR performed with MSCs derived either from non-vitrified (p <0.001) of vitrified (p <0.001) WJ tissue samples, were observed. No statistically significant differences were observed between MSCs derived from non-vitrified and vitrified WJ-tissue samples.

3.7. Indirect Immunofluorescence of HLA-G in WJ-MSCs

Indirect immunofluorescence confirmed the successful expression of HLA-G by MSCs derived either from non-vitrified and vitrified WJ-tissue samples (Figure 6 and Figure S6). No alteration in staining signal was observed between MSCs obtained from both experimental conditions.

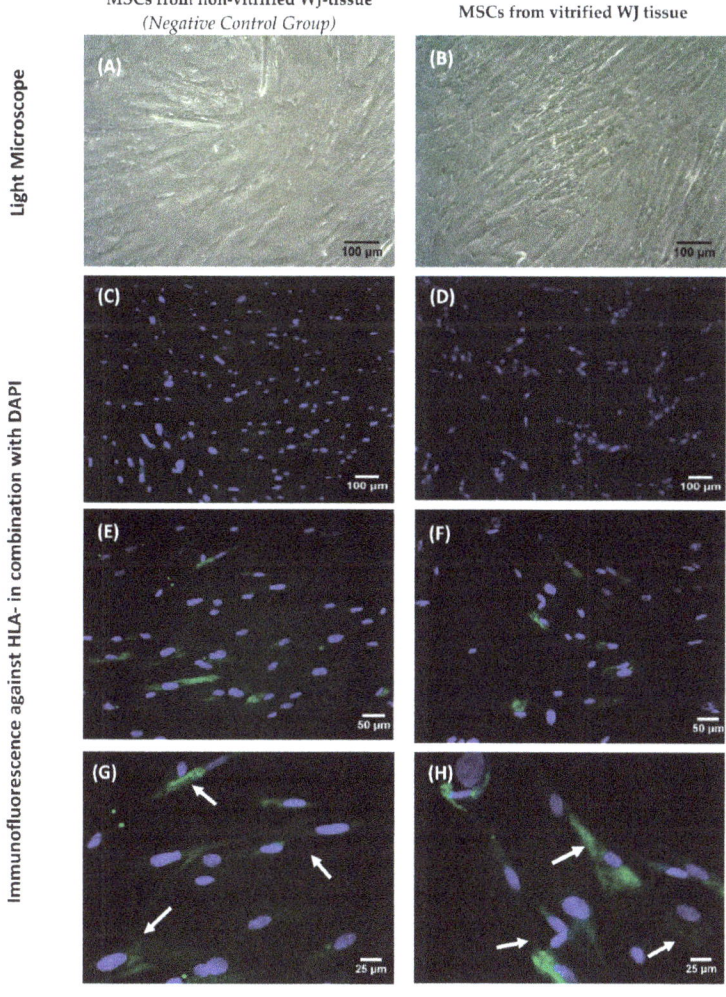

Figure 6. HLA-G expression in WJ-MSCs. WJ-MSCs obtained from non-vitrified (n = 5) and vitrified (n = 5) WJ tissue with light microscopy (**A,B**). Indirect immunofluorescence against HLA-G in combination with DAPI in MSCs derived from non-vitrified (**C,E,G**) and vitrified (**D,F,H**) WJ tissue samples. Images **A,B** were acquired with original magnification 10×, scale bars 100 μm. Images **C,D** were acquired with original magnification 10×, scale bars 100 μm. Images **E,F** were acquired with original magnification 20×, scale bars 50 μm. Images **G,H** were acquired with original magnification 40×, scale bars 25 μm.

4. Discussion

The therapeutic applications of human MSCs towards serious life-threating disorders have been highlighted by several reports [7–9]. MSCs are well known for their immunomodulatory properties, which could make them ideal candidates for personalized medicine. HLA-G seems to a play crucial role in the immunosuppression process. MSCs derived from extraembryonic tissues may be characterized by higher expression of HLA-G than MSCs from other sources [10].

Under this scope, the WJ MSCs could possibly be used in several therapeutic strategies such as administration of GVHD and autoimmune disorders. In most of the times, prolonging cell culture and expansion is needed in order to obtain sufficient number of cells, which could affect significantly the MSCs properties. A possible solution to address this problem, could be the cryopreservation by vitrification of the entire WJ tissue and isolation of MSCs at any desired time point.

The aim of this study was to evaluate the HLA-G expression in MSCs derived from vitrified WJ tissue after long-term storage at $-196\ °C$. Initially, WJ tissues were vitrified and stored for a time period of 1 year at $-196\ °C$. Then, the tissues were thawed and MSCs were isolated. Non-vitrified and CPA-free WJ tissues stored in liquid nitrogen, were used as control groups for this study. MSCs were successfully isolated from vitrified WJ tissue and confluency observed after 18 days, in a similar way as the MSCs derived from non-vitrified WJ tissues. On the other hand, no cells were obtained from CPA-free samples. Moreover, MSCs from non-vitrified and vitrified WJ tissues characterized by the same morphology, while total cell number, CDT, PD and cell viability did not present any statistically significant difference. Furthermore, isolated MSCs from both experimental procedures were successfully passaged and reached passage 8.

MSCs isolated from non-vitrified and vitrified WJ tissue samples were differentiated successfully to "osteocytes", "adipocytes", and "chondrocytes". Furthermore, MSCs formed CFUs, were positive for CD73, CD90, CD105 and negative for CD34, CD45, HLA-DR as indicated by the ISCT [6]. Same results have been observed in several studies, elucidating in this way the efficient storage of WJ tissue over a long time period [15,16]. Moreover, histological analysis revealed that non-vitrified and vitrified WJ tissues were characterized by a dense ECM with well-preserved MSCs, while CPA-free samples exhibited extensive ECM damage. As a consequence to this, the tissue's resident cells were not preserved properly, were damaged and no cells were isolated from CPA-free samples. This phenomenon could be explained by ice crystal formation during storage and thawing procedure of CPA-free samples [14]. On the other hand, the use of a proper combination of cryoprotective agents (low and high molecular weight) in vitrification method, resulted in the preservation of the tissue's ECM and resident cells [14].

Once the proper storage of WJ-tissue at low temperatures was established, the expression of HLA-G was evaluated. HLA-G is expressed primarily in trophoblast and other extraembryonic tissues such as the umbilical cord. Moreover, HLA-G is elevated during pregnancy, thus maintaining in this way the immunosuppressive state towards the fetus. Moreover, the elevation of HLA-G expression is relevant with the successful implantation of trophoblast [21,22]. WJ tissue contains MSCs, which are part of the umbilical cord, thus expressing the HLA-G. Immunohistochemistry results showed the positive expression of HLA-G in WJ tissue from both experimental conditions. In order to evaluate thoroughly that MSCs from both experimental procedures expressed the HLA-G, gene expression analysis was performed.

MSCs derived either from non-vitrified or vitrified WJ-tissue successfully expressed HLA-G1, G5 and G7 isoforms. These results were in accordance with the study of Ding et al. where similar expression levels of HLA-G in MSCs derived from human umbilical cord were observed [20]. HLA-G1 is the membrane bound isoform, which is responsible for prevention of dendritic cell maturation [20]. In addition, HLA-G5 and G7 are the secreted isoforms that are implicated in immune tolerance and allograft acceptance [20]. The ability of MSCs to express both membrane-bound and soluble secreted HLA-G isoforms is very important, making them potential cell populations for treatment of several serious diseases where immune regulation is needed. Moreover, flow cytometric analysis for HLA-G showed that MSCs from both experimental procedures expressed the HLA-G at over 95%, further confirming the initial results from gene expression analysis.

The ability of MSCs to suppress immune cells was checked by an MLR assay. MSCs isolated from non-vitrified and vitrified WJ tissue samples achieved immunosuppression by decreasing the number of responder cells. On the other hand, an increase in the number of responder cells was observed when they interacted with stimulator cells without the presence of MSCs. IFN-γ, which is produced by PMNCs, is responsible for the activation of MSCs. In response to high levels of IFN-γ, MSCs express Intracellular Adhesion Molecule-1 (ICAM-1) and various immunosuppressive factors such as IDO, HLA-G and IL-10. As a result, different immune reactions can be performed either by activation of Th1 or Th2 cells. Furthermore, MSCs can efficiently modulate the immune response by activating T regulatory cells. In this study, where allorecognition was performed, MSCs were capable of suppressing the immune reaction by decreasing the number of responder PBMNCs. Finally, indirect immunofluorescence against HLA-G was performed. These results showed the positive expression of HLA-G in MSCs derived either from non-vitrified or vitrified WJ tissues.

The immunological properties of MSCs presented in the current study seemed to be consistent with previous published studies, where MSCs with different extraembryonic origins were evaluated [20,23–25]. In addition, several reports have focused on the evaluation of immunomodulatory properties of MSCs derived from fresh WJ tissue [20,24,25]. In our study, an initial attempt to evaluate the HLA-G expression in MSCs obtained from vitrified WJ tissue was performed. It is widely known that HLA-G is an immunomodulatory molecule which can interact with tyrosine-based immunoreceptors such as Ig-like transcript 2 (ILT2) and 4 (ILT4) and killer Ig-like receptor (KIR) 2DL4/CD158d [26–28]. Through this interaction, recruitment of Src homology 2 domain-containing tyrosine phosphatase 1 (SHP-1) and 2 (SHP-2), followed by inactivation of Protein Kinase B (PKB) signaling pathway, resulted in cell cycle inactivation [26–29]. In addition, HLA-G can induce T and B lymphocyte apoptosis and the activation of $CD4^+CD25^+FoxP3^+$ regulatory T cells [18]. Due to these immunomodulatory properties, MSCs are an ideal cell population for the administration of GVHD and autoimmune disorders. GVHD and autoimmune disorders are characterized by an extensive immune reaction, where dendritic, T and B cells play crucial roles. As a first line treatment of those patients, is the use of corticosteroids. However, there are patients who develop severe or steroid-refractory GVHD or cannot respond properly to corticosteroid treatment [30–32]. A treatment option might be the infusion of related or unrelated MSCs. Due to their immunomodulatory properties, MSCs can induce tolerance or immune suppression to the patients, avoiding in this way the morbidity and mortality which can be caused. Future experiments will involve the use of MSCs derived from vitrified WJ tissue in animal models with occurred GVHD or autoimmune disorders [30–32]. In addition, the expression of HLA-G can be evaluated between MSCs from different sources such as bone marrow and adipose tissue in order to thoroughly assess their immunomodulatory properties.

5. Conclusions

MSCs derived from vitrified WJ tissue can efficiently retain their HLA-G expression. Cryopreservation by vitrification can be used for the proper storage of WJ tissue over a long-time period. MSCs can be isolated from vitrified WJ tissues and expanded successfully under GMP conditions, thus yielding a great number of cells that could be used in personalized medicine approaches.

Supplementary Materials: The following are available online at http://www.mdpi.com/2306-5354/5/4/95/s1, Figure S1: Days of first evidence of isolated MSCs from non-vitrified, vitrified and CPA-free WJ tissue samples. Figure S2: Initial passage of MSCs to cell culture flasks. Figure S3: Passages of MSCs derived from non-vitrified and vitrified WJ tissue samples from 1 to 8. Table S1: Flow cytometric analysis of HLA-G expression in WJ -MSCs. Table S2: MLR results. Figure S4: Mixed Lymphocyte reaction. Figure S5: HLA-G expression in WJ tissue. Figure S6: Indirect immunofluorescence for HLA-G expression in WJ-MSCs.

Author Contributions: P.M. (first author) and D.B. carried out the whole experimental procedure of this study. In addition, P.M. performed the statistical analysis. A.D. and M.S.-V. validated the HLA-G expression results. E.M. supervised the study. C.S.-G. supervised and approved the overall study.

Funding: This research received no funding.

Conflicts of Interest: The authors declare no conflict of interest.

References

1. Fitzsimmons, R.E.B.; Mazurek, M.S.; Soos, A.; Simmons, C.A. Mesenchymal Stromal/Stem Cells in Regenerative Medicine and Tissue Engineering. *Stem Cells Int.* **2018**, *19*, 8031718. [CrossRef] [PubMed]
2. Majka, M.; Sułkowski, M.; Badyra, B.; Musiałek, P. Concise Review: Mesenchymal Stem Cells in Cardiovascular Regeneration: Emerging Research Directions and Clinical Applications. *Stem Cells Transl. Med.* **2017**, *6*, 1859–1867. [CrossRef] [PubMed]
3. Jin, Y.Z.; Lee, J.H. Mesenchymal Stem Cell Therapy for Bone Regeneration. *Clin. Orthop. Surg.* **2018**, *10*, 271–278. [CrossRef] [PubMed]
4. Abumaree, M.H.; Abomaray, F.M.; Alshabibi, M.A.; AlAskar, A.S.; Kalionis, B. Immunomodulatory properties of human placental mesenchymal stem/stromal cells. *Placenta* **2017**, *59*, 87–95. [CrossRef] [PubMed]
5. Wang, L.T.; Ting, C.H.; Yen, M.L.; Liu, K.J.; Sytwu, H.K.; Wu, K.K.; Yen, B.L. Human mesenchymal stem cells (MSCs) for treatment towards immune- and inflammation-mediated diseases: Review of current clinical trials. *J. Biomed. Sci.* **2016**, *23*, 76. [CrossRef] [PubMed]
6. Dominici, M.; Le Blanc, K.; Mueller, I.; Slaper-Cortenbach, I.; Marini, F.C.; Krause, D.S.; Deans, R.J.; Keating, A.; Prockop, D.J.; Horwitz, E.M. Minimal criteria for defining multipotent mesenchymal stromal cells. The international society for cellular therapy position statement. *Cytotherapy* **2006**, *8*, 315–317. [CrossRef] [PubMed]
7. Wang, L.; Zhu, C.Y.; Ma, D.X.; Gu, Z.Y.; Xu, C.C.; Wang, F.Y.; Chen, J.G.; Liu, C.J.; Guan, L.X.; Gao, R.; et al. Efficacy and safety of mesenchymal stromal cells for the prophylaxis of chronic graft-versus-host disease after allogeneic hematopoietic stem cell transplantation: A meta-analysis of randomized controlled trials. *Ann. Hematol.* **2018**, *97*, 1941–1950. [CrossRef] [PubMed]
8. Servais, S.; Baron, F.; Lechanteur, C.; Seidel, L.; Selleslag, D.; Maertens, J.; Baudoux, E.; Zachee, P.; Van Gelder, M.; Noens, L.; et al. Infusion of bone marrow derived multipotent mesenchymal stromal cells for the treatment of steroid-refractory acute graft-versus-host disease: A multicenter prospective study. *Oncotarget* **2018**, *9*, 20590–20604. [CrossRef] [PubMed]
9. De Araújo Farias, V.; Carrillo-Gálvez, A.B.; Martín, F.; Anderson, P. TGF-β and mesenchymal stromal cells in regenerative medicine, autoimmunity and cancer. *Cytokine Growth Factor Rev.* **2018**, *43*, 25–37. [CrossRef] [PubMed]
10. Naji, A.; Rouas-Freiss, N.; Durrbach, A.; Carosella, E.D.; Sensébé, L.; Deschaseaux, F. Concise review: Combining human leukocyte antigen G and mesenchymal stem cells for immunosuppressant biotherapy. *Stem Cells* **2013**, *31*, 2296–2303. [CrossRef] [PubMed]
11. Selmani, Z.; Naji, A.; Zidi, I.; Favier, B.; Gaiffe, E.; Obert, L.; Borg, C.; Saas, P.; Tiberghien, P.; Rouas-Freiss, N.; et al. Human leukocyte antigen-G5 secretion by human mesenchymal stem cells is required to suppress T lymphocyte and natural killer function and to induce $CD4^+CD25^{high}FOXP3^+$ regulatory T cells. *Stem Cells* **2008**, *26*, 212–222. [CrossRef] [PubMed]
12. Bakopoulou, A.; Apatzidou, D.; Aggelidou, E.; Gousopoulou, E.; Leyhausen, G.; Volk, J.; Kritis, A.; Koidis, P.; Geurtsen, W. Isolation and prolonged expansion of oral mesenchymal stem cells under clinical-grade, GMP-compliant conditions differentially affects "stemness" properties. *Stem Cell Res. Ther.* **2017**, *8*, 247. [CrossRef] [PubMed]
13. Chatzistamatiou, T.K.; Papassavas, A.C.; Michalopoulos, E.; Gamaloutsos, C.; Mallis, P.; Gontika, I.; Panagouli, E.; Koussoulakos, S.L.; Stavropoulos-Giokas, C. Optimizing isolation culture and freezing methods to preserve Wharton's jelly's mesenchymal stem cell (MSC) properties: An MSC banking protocol validation for the Hellenic Cord Blood Bank. *Transfusion* **2014**, *54*, 3108–3120. [CrossRef] [PubMed]
14. Fahy, G.M.; Wowk, B. Principles of cryopreservation by vitrification. *Methods Mol. Biol.* **2015**, *1257*, 21–82. [PubMed]
15. Shivakumar, S.B.; Bharti, D.; Subbarao, R.B.; Jang, S.J.; Park, J.S.; Ullah, I.; Park, J.K.; Byun, J.H.; Park, B.W.; Rho, G.J. DMSO- and Serum-Free Cryopreservation of Wharton's Jelly Tissue Isolated from Human Umbilical Cord. *J. Cell. Biochem.* **2016**, *117*, 2397–2412. [CrossRef] [PubMed]

16. Fong, C.Y.; Subramanian, A.; Biswas, A.; Bongso, A. Freezing of Fresh Wharton's Jelly from Human Umbilical Cords Yields High Post-Thaw Mesenchymal Stem Cell Numbers for Cell-Based Therapies. *J. Cell. Biochem.* **2016**, *117*, 815–827. [CrossRef] [PubMed]
17. Paladino, F.V.; Sardinha, L.R.; Piccinato, C.A.; Goldberg, A.C. Intrinsic Variability Present in Wharton's Jelly Mesenchymal Stem Cells and T Cell Responses May Impact Cell Therapy. *Stem Cells Int.* **2017**, *2017*, 8492797. [CrossRef] [PubMed]
18. Wang, Q.; Yang, Q.; Wang, Z.; Tong, H.; Ma, L.; Zhang, Y.; Shan, F.; Meng, Y.; Yuan, Z. Comparative analysis of human mesenchymal stem cells from fetal-bone marrow, adipose tissue, and Warton's jelly as sources of cell immunomodulatory therapy. *Hum. Vaccin Immunother.* **2016**, *12*, 85–96. [CrossRef] [PubMed]
19. Mattar, P.; Bieback, K. Comparing the Immunomodulatory Properties of Bone Marrow, Adipose Tissue, and Birth-Associated Tissue Mesenchymal Stromal Cells. *Front. Immunol.* **2015**, *6*, 560. [CrossRef] [PubMed]
20. Ding, D.C.; Chou, H.L.; Chang, Y.H.; Hung, W.T.; Liu, H.W.; Chu, T.Y. Characterization of HLA-G and Related Immunosuppressive Effects in Human Umbilical Cord Stroma-Derived Stem Cells. *Cell Transplant.* **2016**, *25*, 217–228. [CrossRef] [PubMed]
21. Fanchin, R.; Gallot, V.; Rouas-Freiss, N.; Frydman, R.; Carosella, E.D. Implication of HLA-G in human embryo implantation. *Hum. Immunol.* **2007**, *68*, 259–263. [CrossRef] [PubMed]
22. Steinborn, A.; Varkonyi, T.; Scharf, A.; Bahlmann, F.; Klee, A.; Sohn, C. Early detection of decreased soluble HLA-G levels in the maternal circulation predicts the occurrence of preeclampsia and intrauterine growth retardation during further course of pregnancy. *Am. J. Reprod. Immunol.* **2007**, *57*, 277–286. [CrossRef] [PubMed]
23. Selmani, Z.; Naji, A.; Gaiffe, E.; Obert, L.; Tiberghien, P.; Rouas-Freiss, N.; Carosella, E.D.; Deschaseaux, F. HLA-G is a crucial immunosuppressive molecule secreted by adult human mesenchymal stem cells. *Transplantation* **2009**, *87* (Suppl. 9), S62–S66. [CrossRef] [PubMed]
24. Stubbendorff, M.; Deuse, T.; Hua, X.; Phan, T.T.; Bieback, K.; Atkinson, K.; Eiermann, T.H.; Velden, J.; Schröder, C.; Reichenspurner, H.; et al. Immunological properties of extraembryonic human mesenchymal stromal cells derived from gestational tissue. *Stem Cells Dev.* **2013**, *22*, 2619–2629. [CrossRef] [PubMed]
25. Kita, K.; Gauglitz, G.G.; Phan, T.T.; Herndon, D.N.; Jeschke, M.G. Isolation and characterization of mesenchymal stem cells from the sub-amniotic human umbilical cord lining membrane. *Stem Cells Dev.* **2010**, *19*, 491–502. [CrossRef] [PubMed]
26. La Rocca, G.; Anzalone, R.; Corrao, S.; Magno, F.; Loria, T.; Lo Iacono, M.; Di Stefano, A.; Giannuzzi, P.; Marasà, L.; Cappello, F.; et al. Isolation and characterization of Oct-4$^+$/HLA-G$^+$ mesenchymal stem cells from human umbilical cord matrix: Differentiation potential and detection of new markers. *Histochem. Cell Biol.* **2009**, *131*, 267–282. [CrossRef] [PubMed]
27. Carosella, E.D.; Favier, B.; Rouas-Freiss, N.; Moreau, P.; Lemaoult, J. Beyond the increasing complexity of the immunomodulatory HLA-G molecule. *Blood* **2008**, *111*, 4862–4870. [CrossRef] [PubMed]
28. Rouas-Freiss, N.; Khalil-Daher, I.; Riteau, B.; Menier, C.; Paul, P.; Dausset, J.; Carosella, E.D. The immunotolerance role of HLA-G. *Semin. Cancer Biol.* **1999**, *9*, 3–12. [CrossRef] [PubMed]
29. Gregori, S.; Tomasoni, D.; Pacciani, V.; Scirpoli, M.; Battaglia, M.; Magnani, C.F.; Hauben, E.; Roncarolo, M.G. Differentiation of type 1 T regulatory cells (Tr1) by tolerogenic DC-10 requires the IL-10-dependent ILT4/HLA-G pathway. *Blood* **2010**, *116*, 935–944. [CrossRef] [PubMed]
30. Goto, T.; Murata, M.; Terakura, S.; Nishida, T.; Adachi, Y.; Ushijima, Y.; Shimada, K.; Ishikawa, Y.; Hayakawa, F.; Nishio, N.; et al. Phase I study of cord blood transplantation with intrabone marrow injection of mesenchymal stem cells: A clinical study protocol. *Medicine* **2018**, *97*, e0449. [CrossRef] [PubMed]
31. Liu, Z.; Zhang, Y.; Xiao, H.; Yao, Z.; Zhang, H.; Liu, Q.; Wu, B.; Nie, D.; Li, Y.; Pang, Y.; et al. Cotransplantation of bone marrow-derived mesenchymal stem cells in haploidentical hematopoietic stem cell transplantation in patients with severe aplastic anemia: An interim summary for a multicenter phase II trial results. *Bone Marrow Transplant.* **2017**, *52*, 1080. [CrossRef] [PubMed]
32. Fan, J.; Tang, X.; Wang, Q.; Zhang, Z.; Wu, S.; Li, W.; Liu, S.; Yao, G.; Chen, H.; Sun, L. Mesenchymal stem cells alleviate experimental autoimmune cholangitis through immunosuppression and cytoprotective function mediated by galectin-9. *Stem Cell Res Ther.* **2018**, *9*, 237. [CrossRef] [PubMed]

© 2018 by the authors. Licensee MDPI, Basel, Switzerland. This article is an open access article distributed under the terms and conditions of the Creative Commons Attribution (CC BY) license (http://creativecommons.org/licenses/by/4.0/).

MDPI
St. Alban-Anlage 66
4052 Basel
Switzerland
Tel. +41 61 683 77 34
Fax +41 61 302 89 18
www.mdpi.com

Bioengineering Editorial Office
E-mail: bioengineering@mdpi.com
www.mdpi.com/journal/bioengineering

www.ingramcontent.com/pod-product-compliance
Lightning Source LLC
LaVergne TN
LVHW071959080526
838202LV00064B/6793

9 783039 214976